Emulsions, Latices, and Dispersions

Emulsions, Latices, and Dispersions

Edited by

PAUL BECHER
ICI Americas Inc.
Wilmington, Delaware

•

MARVIN N. YUDENFREUND
Drew Chemical Corporation
Boonton, New Jersey

MARCEL DEKKER, INC. New York and Basel

Library of Congress Cataloging in Publication Data

Main entry under title:

Emulsions, latices, and dispersions.

Selection of papers presented at a symposium held as part of the 51st Colloid and Surface Science Symposium, held at Grand Isle, N. Y. in June 1977.
Includes bibliographical references and index.
1. Colloids--Congresses. 2. Emulsions--Congresses.
I. Becher, Paul. II. Yudenfreund, Marvin N., [Date]
QD549.E49 541'.3451 78-11192
ISBN 0-8247-6797-7

270 Madison Avenue, New York, New York 10016

Current Printing (last digit):
10 9 8 7 6 5 4 3 2 1

PRINTED IN THE UNITED STATES OF AMERICA

PREFACE

The papers in this collection were presented as part of a session on "Emulsions, Dispersions, and Latices" at the 51st Colloid and Surface Science Symposium of the Division of Colloid and Surface Chemistry, held at Grand Isle, N.Y., in June of 1977. This collection consists of approximately half the papers presented at the Symposium; in the case of most of the papers not represented, prior commitments as to publication precluded their inclusion here.

The theme of the session was stated broadly, so that almost any aspect of disperse systems could be included. This was deliberate. Rather than -- as some might think -- serving to dilute the impact of the proceedings, it illustrated forcefully the way in which the general theories of colloid stability cut across disciplinary borders. "All chemistry," said Wilder D. Bancroft, "is colloid chemistry," and the truth of that dictum is illustrated by the present collection.

It is equally well-illustrated by the one paper included in this volume which was not delivered at the 51st Symposium as part of the session, but rather as a plenary lecture. We are most fortunate to have persuaded Professor J. Th. G. Overbeek to allow us to include his plenary lecture, which so brilliantly expresses what might be called Bancroft's Other Rule.

The editors wish to express their gratitude to the organizers of the 51st Symposium, and especially to the Chairman, Professor Robert J. Good, for giving us the opportunity of organizing the program, and thus, in a sense, acting as godparents to this book. The responsibility for any faults, however, is exclusively that of the editors.

Paul Becher
Wilmington, Delaware

Marvin N. Yudenfreund
Boonton, New Jersey

CONTRIBUTORS

Syed M. Ahmed
Departments of Chemical Engineering and Chemistry and Emulsion Polymers Institute, Lehigh University, Bethlehem, Pennsylvania

C. Boned
Laboratoire de Thermodynamique, Institut Universitaire de Recherche Scientifique, Université de Pau et des Pays de l'Adour, Pau, France

*John P. Carrera**
Department of Chemical Engineering, Carnegie-Mellon University, Pittsburgh, Pennsylvania

Marc Clausse
Laboratoire de Thermodynamique, Institut Universitaire de Recherche Scientifique, Université de Pau et des Pays de l'Adour, Pau, France

C. A. Daniels
Avon Lake Technical Center, BFGoodrich Chemical Division, Avon Lake, Ohio

J. A. Davidson
Avon Lake Technical Center, BFGoodrich Chemical Division, Avon Lake, Ohio

Mohamed S. El-Aasser
Emulsion Polymers Institute and Departments of Chemical Engineering, Lehigh University, Bethlehem, Pennsylvania

Leif Eriksson
The Swedish Institute for Surface Chemistry, Stockholm, Sweden

Norman Epstein
Chemical Engineering Department, University of British Columbia, Vancouver, Canada

Ronald Flaska
Departments of Chemical Engineering and Chemistry and Emulsion Polymers Institute, Lehigh University, Bethlehem, Pennsylvania

---
*Currently with Exxon Research and Engineering Co., Baytown, Texas

J. R. Ford
Department of Chemistry, University of Massachusetts, Amherst, Massachusetts

Stig Friberg
Department of Chemistry, University of Missouri-Rolla, Rolla, Missouri

Gunilla Gillberg
Department of Chemistry, University of Missouri-Rolla, Rolla, Missouri

Yuli Glazman
Department of Chemical Engineering, Tufts University, Medford, Massachusetts

Tamotsu Kondo
Faculty of Pharmaceutical Sciences, Science University of Tokyo, Tokyo, Japan

Irvin M. Krieger
Departments of Macromolecular Science and Chemistry, Case Western Reserve University, Cleveland, Ohio

R. J. Kuo
Department of Chemistry, Kent State University, Kent, Ohio

*Samuel Levine**
Chemical Engineering Department, University of British Columbia, Vancouver, Canada

Li-Jen Liu
Department of Chemistry, Case Western Reserve University, Cleveland, Ohio

S. A. McDonald
Avon Lake Technical Center, BFGoodrich Chemical Division, Avon Lake, Ohio

Anthony J. McHugh
Department of Chemical Engineering and Emulsion Polymers Institute, Lehigh University, Bethlehem, Pennsylvania

F. J. Micale
Emulsion Polymers Institute, Lehigh University, Bethlehem, Pennsylvania

*Permanent Address: Mathematics Department, University of Manchester, Manchester, England

R. R. Myers
Department of Chemistry, Kent State University, Kent, Ohio

Graham Neale
Chemical Engineering Department, University of Ottawa, Ottawa, Canada

Gary Nishioka
Department of Chemistry, Rensselaer Polytechnic Institute, Troy, New York

J. Th. G. Overbeek
Van 't Hoff Laboratory for Physical and Colloid Chemistry, University of Utrecht, Utrecht, The Netherlands

J. W. Parsons
Department of Chemistry, University of Massachusetts, Amherst, Massachusetts

J. Peyrelasse
Laboratoire de Thermodynamique, Institut Universitaire de Recherche Scientifique, Université de Pau et des Pays de l'Adour, Pau, France

Gary W. Poehlein
Emulsion Polymers Institute and Departments of Chemical Engineering and Chemistry, Lehigh University, Bethlehem, Pennsylvania

Dennis C. Prieve
Department of Chemical Engineering, Carnegie-Mellon University, Schenley Park, Pittsburgh, Pennsylvania

Sidney Ross
Department of Chemistry, Rennselaer Polytechnic Institute, Troy, New York

R. L. Rowell
Department of Chemistry, University of Massachusetts, Amherst, Massachusetts

R. J. Ruch
Department of Chemistry, Kent State University, Kent, Ohio

Richard L. Schild
Emulsion Polymers Institute, Lehigh University, Bethlehem, Pennsylvania

Cesar A. Silebi
Department of Chemical Engineering and Emulsion Polymers Institute, Lehigh University, Bethlehem, Pennsylvania

Motoharu Shiba
Faculty of Pharmaceutical Sciences, Science University of Tokyo, Tokyo, Japan

Akemi Tateno
Faculty of Pharmaceutical Sciences, Science University of Tokyo, Tokyo, Japan

John W. Vanderhoff
Departments of Chemical Engineering and Chemistry and Emulsion Polymers Institute, Lehigh University, Bethlehem, Pennsylvania

S. R. Vasconcellos
Department of Chemistry, University of Massachusetts, Amherst, Massachusetts

P. Xans
Laboratoire de Thermodynamique, Institut Universitaire de Recherche Scientifique, Université de Pau et des Pays de l'Adour, Pau, France

W. C. Wu
Emulsion Polymers Institute, Lehigh University, Bethlehem, Pennsylvania

CONTENTS

Emulsions and Dispersions

Emulsions, Latices, and Dispersions

TECHNICAL APPLICATIONS OF COLLOID SCIENCE

J.Th.G. Overbeek

Van 't Hoff Laboratory for Physical and Colloid Chemistry
University of Utrecht
Utrecht, The Netherlands

ABSTRACT

Colloidal dimensions play a role in many applications. In preparing dispersed systems, condensation methods lead to finer and more uniform dispersions than dispersion methods. In technical preparation of dispersions a two step condensation-dispersion procedure is frequently followed. Among the examples treated in more detail are emulsion polymerization leading to isodispersed latices, and "throw-away dispersions" as encountered in washing. In applications often a state of thixotropy or weak flocculation is desirable. Special attention is given to painting, electro-deposition and drilling muds. In some applications coagulation or coalescence is essential e.g. in natural crude oil emulsions, in coal washeries, using sensitized flocculation and in froth flotation.

I. INTRODUCTION

I have chosen this subject for my lecture in order to be able to stress once more, in how many, often quite dissimilar, fields colloid science is applied, or rather in how many applications colloidal dimensions and colloidal phenomena are important. Also for those who occupy themselves with pure research and with education it is essential to be aware of the broad applicability of colloid and surface science.

Some obvious cases where the *dispersed state is desirable* are: milk, latex, many paints, inks and laquers, mayonnaise, photographic "emulsions", dirt suspended by soap, but also soap micelles as such, and solutions of polymers and biopolymers. Living organisms make and use dispersed systems. Blood is an outstanding example. One may consider the individual cell itself to be a colloidal particle and wonder why cells usually have dimensions between 1 and 10 μm, and why elephants are not unicellular. The answer, of course, is that diffusion is a quick enough transport mechanism in water over distances of the order of one micron but much too slow for larger distances. That is why larger (and higher) organisms are multicellular and why they need circulation systems. Many organisms grow colloidal fibers. They have been used by man and have stood model for a very large industry.

Most of the above examples deal with dispersions in liquid. Dispersions in gases, aerosols, are also applied, e.g. for camouflage, for protection of orchards against frost damage by radiation, and in fluidized bed reactors.

The *dispersed state*, however, is *not always desirable*. In most macroscopic phase separations, in liquid-liquid extractions, in water and air purification by sedimentation and/or filtration, a stable state of fine dispersion is a drawback. In the preparation of polymers by emulsion polymerization, in latex based paints, the colloidal state, essential in the early stages, has to be destroyed towards the end of the process. Foams and emulsions are often a nuisance. They have to be broken or their formation has to be prevented.

I could continue giving other examples but I will rather select a relatively small number of typical applications and treat these a little more in depth.

II. PREPARATION OF STABLE DISPERSIONS

It is customary to divide methods for preparation of colloidal dispersions into *dispersion methods* and *condensation methods* [1]. In the dispersion methods coarse material is subjected to milling, grinding or other methods of comminution, until the desired degree of fineness is reached. The presence of a stabilizing agent is essential to prevent the fine particles from forming agglomerates. The stabilizing agent may be either an ionic species that is sufficiently strongly adsorbed or a large adsorbable molecule or a combination of both (polyion). In the condensation methods the particles are allowed to grow, starting from an atomic, ionic or molecular solution (or from the gasphase) until the desired size is reached. Again a stabilizing agent must be present. Good examples are:

the preparation of a gold sol by the reduction of a chloroaurate solution;

$AuCl_4^-$ + reducer (e.g. CH_2O) $\rightarrow$ Au + 4 Cl^- + oxidation product,

and the preparation of a silver iodide sol from solutions of $AgNO_3$ and KI;

$$Ag^+ + I^- \rightarrow AgI.$$

The gold sol is stabilized by adsorption of negative ions (e.g. chlorogold complexes, citrate or OH^-). In the case of the silver iodide a small excess of Ag^+ or I^- provides the stabilizing charge.

It is often stressed that the condensation methods allow finer, more isodispersed, particles to be obtained and most laboratory methods for the preparation of lyophobic sols are based on condensation. In technical applications, however, dispersion methods are more commonly used than condensation methods, with the preparation of photographic emulsions and emulsion polymerization as the outstanding exceptions. Dispersed paints and dyes, most pharmaceutical, cosmetic and agricultural emulsions and suspensions, suspensions used in froth flotation, drilling muds, fluorescent light tubes, magnetic tapes and composite multiphase materials are all made via dispersion methods.

This, however, does not imply that in the dispersion methods one starts with very coarse bulk crystals. No, the material to be milled is often prepared by a condensation method from low molecular weight starting materials, but the condensation is allowed to go on beyond the degree of subdivision eventually required. The reason for such a two step, condensation-dispersion, procedure is that after the condensation, ageing and thermal treatments give the necessary control of magnetic properties, color or fluorescence of the particles, after which the final dispersion is prepared by milling in the presence of the stabilizing agent.

Examples

1. PHOTOGRAPHIC EMULSIONS

As mentioned above *photographic emulsions* [2] (an unfortunate misnomer, given the suspension character of the system) are formed by condensation. Solutions of silvernitrate and alkali halide in gelatin are mixed and a very fine dispersion of silver halide is formed. The gelatin serves both as the stabilizing agent (protective colloid) and as the gel that immobilizes the system.

Washing away of the alkali nitrate, addition of sensitizer and dyes and carefully controlled Ostwald ripening to obtain good sensitivity are essential steps in the process.

2. EMULSION POLYMERIZATION

In *emulsion polymerization* [3] the latex particles are formed by condensation starting from molecularly distributed monomer(s). The monomers may originally be present as relatively coarse emulsion droplets, or be dissolved in water. The essential first steps, the formation of free radicals from the initiator and the addition of the first few monomers take place in the aqueous phase. The growing polymer becomes insoluble, separates from the solution and forms the latex particles. The size of these particles is determined by the number of latex nuclei formed and the time during which they grow. An isodisperse latex may be expected when all the nuclei are formed at a very early stage, or if instead of allowing spontaneous nucleation the system is seeded with a large number of small latex particles. Agglomeration must of course be prevented by sufficiently strong stabilization either by the action of suitable soaps or other amphipathic substances or by the electric charge on the polymers themselves.

It may not be superfluous to point out why *isodispersed systems* [4] are obtained, if all nucleation takes place at an early stage of the process. Growth of the nuclei requires diffusion of the monomers to the latex particles (or diffusion of other low molecular weight building stones in other cases such as Au atoms to the growing particles in a gold sol), followed by addition of monomers to the growing polymer molecules (or precipitation, crystallization of the small building blocks on the growing particle).

Each of these steps may be rate limiting.

a. If the diffusion is rate limiting the flux of small molecules, J, towards each particle can be found, using a reasoning analogous to Smoluchowski's derivation [5] of the rate of coagulation, which also applies to diffusion controlled reactions [6]. A concentration gradient which soon reaches a stationary state, develops around each particle. Assuming spherical symmetry this flux in the stationary state is

$$J = D\, 4\pi r^2\, dc/dr \qquad (1)$$

where c is the concentration of the small molecules, r the distance from the center of the particle and D the mutual diffusion coefficient between particle and small molecules. D is equal to

$$D = D_p + D_m = D_m\ (1 + \frac{b}{a}) \qquad (2)$$

p and m refer to particle and monomer (small molecules) respectively, b is the radius of the monomer and a that of the particle. Since the diffusion is considered to be rate limiting the concentration c at the particle surface, c_a, must be zero or at least very low. If the diffusion is not rate limiting, c_a must be equal to the monomer activity in the swollen particle. It will be constant or slowly changing with time and may have any value between zero and the concentration far from the particle, which we shall call c_∞. Integrating eq. (1) we find

$$J = 4\pi Da\ (c_\infty - c_a) \simeq 4\pi Dac_\infty \qquad (3)$$

The rate of growth of the particle radius a is then

$$\frac{da}{dt} = \frac{J}{4\pi a^2}\bar{V} = D_m c_\infty \bar{V}\ (\frac{1}{a} + \frac{b}{a^2}) \qquad (4)$$

where $\bar{V}$ is the partial molar volume of the polymerized monomer in the particle.

If the growth is not very fast (e.g. $da/dt \leqslant 1$ Å s^{-1}) and realistic values for D_m, $\bar{V}$ and a are chosen, c_∞ must be extremely small (e.g. 10^{-9} g cm^{-3}). Then, however, the supply of m would be soon exhausted and reasonable concentrations of particles of acceptable size can only be obtained if the low concentration, c_∞, is continuously replenished by slow production of m [4].

Integration of eq. (4) leads to

$$\tfrac{1}{2}(a^2-a_o^2) - b(a-a_o) + b^2 \ln \frac{a+b}{a_o+b} = D_m c_\infty \bar{V} t \qquad (5)$$

After a given time t the variation in radius of the particle da can be compared to the variation da_o at $t = 0$.

This is found to be

$$\frac{da}{da_o} = \frac{a_o}{a} \frac{1 + b/a}{1 + b/a_o} \qquad (6)$$

showing that the distribution of radii becomes narrower with time. On growing, the particles get more and more isodispersed.

b. If, however, the rate of growth of the particles is limited by the rate of polymerization (or crystallization or precipitation) the increase in particle volume with time is

$$\frac{d(4\pi a^3/3)}{dt} = s\bar{V} \qquad (7)$$

where s is the amount of monomer polymerizing per unit time and per particle. s may be proportional to a^3 (number of growing radicals per unit volume is constant) or proportional to a^2 (constant number of growing radicals per unit surface area) or independent of a (constant, presumably low - 1 or 0.5 - number of growing chains per particle).

If s is constant, eq. (7) can be integrated over t and differentiated by a_o. We find then:

$$\frac{da}{da_o} = \frac{a_o^2}{a^2} \qquad (8)$$

The narrowing of the particle size distribution with time is still more pronounced than when diffusion is rate limiting. Similarly proportionality between s and a^2 leads to $da/da_o = 1$ and with growing a the distribution becomes relatively narrower.

Only if s is proportional to a^3, $da/da_o = a/a_o$, with just no relative change in the distribution on growth.

In all these cases it is essential that all particles grow during the same time, which is equivalent to the requirement that all nucleation takes place at a very early stage.

Vanderhoff et al. [7] have pointed out already in 1956 that, as long as the growth rate is proportional to a smaller power than the third of the radius the particle, size distribution is self sharpening. From experiments with a mixture of two polystyrene latices of different size they found the rate of volume growth to be proportional to the radius to the power 2.5. This might be interpreted as due to a constant number of growing chains per unit area, plus some extra ones, but not so many as to make the number proportional to the particle volume.

3. DISPERSED PIGMENTS AND DYESTUFFS

Dispersed pigments and dyestuffs [8] are prepared by milling the coarsely divided solids in the liquid dispersion medium in the presence of a stabilizer. The milling leads to a fairly wide size distribution, often limited at the top side by continuing the milling until the requirement that all (a very large fraction of) the particles are below a given size is satisfied. At the lower end the size is limited because milling of any type becomes less effective when smaller particles must be broken. Since the process is applied to a great variety of solvents (varying from water to extremely non-polar hydrocarbons) and a great variety of solids, a wide choice of stabilizers must be available. A stabilizer must satisfy three requirements, viz.

a. it must be soluble in the medium,
b. it must be adsorbed on the particles and
c. it must cause a sufficiently large repulsion between the particles.

In a number of cases, when the medium is water or a fairly polar organic solvent, small inorganic ions, in particular potential determining ions, or detergent ions satisfy these requirements. In these cases the repulsion is electrostatic in nature. More often than not, electrical stabilization is not practical (ionic strength too high, medium not polar enough) and then stabilizers must contain a fairly large group (often a long chain) which is soluble in the medium and causes so called "steric" or "entropic" repulsion and an anchorgroup, which has a strong tendency to be adsorbed on the particles but which, on the other hand, should not cause the stabilizer to become insoluble. The preparation and selection of such stabilizers is still more a highly developed art than a deductive science.

When polyelectrolytes (among them gums and proteins) are used as stabilizers we probably have cases of mixed electrical and steric stabilization.

In technical applications it is frequently desirable that dispersions are not completely stabilized, but that they do retain a certain thixotropy, which is a sign of weak flocculation. This is particularly obvious in the case of paints. If the pigment particles in a paint are completely stabilized, on standing a very dense sediment will be formed at the bottom of the can, which can only be redispersed with difficulty. A small degree of thixotropy will prevent the sedimentation, but still allow the paint to behave as a freely mobile liquid during the application. The thixotropy also helps to keep the paint film from flowing down a vertical surface in the time between application and drying (hardening).

4. EMULSIONS

Emulsions [9] are applied in many cases. They can be easily prepared by stirring the two liquids together and subjecting the coarse emulsion to high shear, because a sufficiently elongated

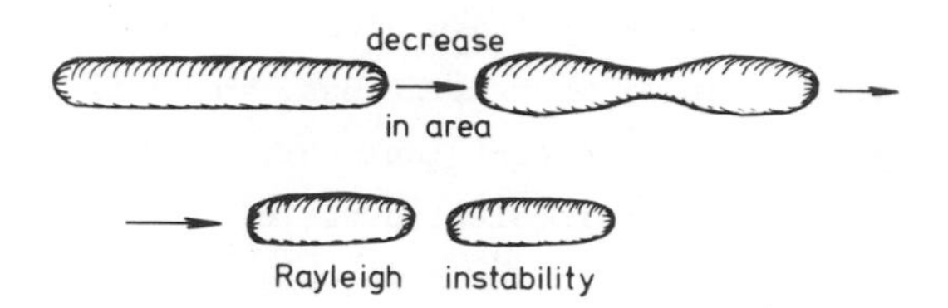

Fig. 1. Illustrating the breaking of an elongated drop by Rayleigh instability.

drop will break spontaneously into two or more smaller ones under decrease of the total interfacial area as illustrated in Fig. 1. A stabilizer is essential to prevent reaggregation (coalescence), and it depends on the nature of the stabilizer and to a lesser extent on the phase volume, whether oil (general designation for a water insoluble organic liquid) in water or water in oil emulsions are obtained. We leave liquid metal emulsions (e.g. Hg in H_2O) out of consideration.

Emulsions prepared by stirring are usually fairly coarse and heterodisperse. Homogenizing produces finer emulsions, but still the droplet diameter does not fall below a few tenths of a micron and the droplet size distribution remains wide. Moreover, making fine emulsions by this method requires large amounts of energy. So called *spontaneous emulsification*, in which emulsions are formed with little or no mechanical energy applied, are therefore often an attractive proposition. A number of mechanisms, leading to fluctuations that grow spontaneously, may contribute individually or collectively to spontaneous emulsification. Diffusion of material across the phase boundary is one of these mechanisms. It may cause density instability and Benard convection cells, which may entrain droplets of one phase into the other. The turbulent motions may be enhanced by changes of interfacial tension along the interface caused by concentration gradients. A temporary, local, negative interfacial tension would also cause the surface area to increase and be a mechanism for spontaneous emulsification.

Certain agricultural chemicals are sold in the form of such *self emulsifiable concentrates*. They consist as a rule of the active substance dissolved in kerosene, to which an oil soluble surfactant or mixture of surfactants, which is also to some extent water soluble, has been added.

A closely related phenomenon is presented by the *micro-emulsions*, formed spontaneously from a mixture of water, oil, a fairly polar surfactant and a higher alcohol or amine. Here, indeed the original two phase mixture may have a negative interfacial tension, but the large interfacial area requires so many surfactant molecules, that their concentration is substantially reduced and the interfacial tension brought back to a positive value. The particle size of microemulsions is very small, of the order of a few hundred Ångstrøms, and they appear to be rather isodisperse.

5. WASHING

In *washing* [10] solid and liquid soil particles should be dispersed and emulsified and be prevented from redeposition during the washing and rinsing process. The washing solution must easily wet the objects (or subjects) to be cleaned. This requires spreading and is promoted by a low surface tension. The stabilizer, in becoming adsorbed, is the cause of a repulsion between soil particle and skin, fabric, glass, ceramic etc. as illustrated in Fig. 2. This repulsion may lead to peptization but often some

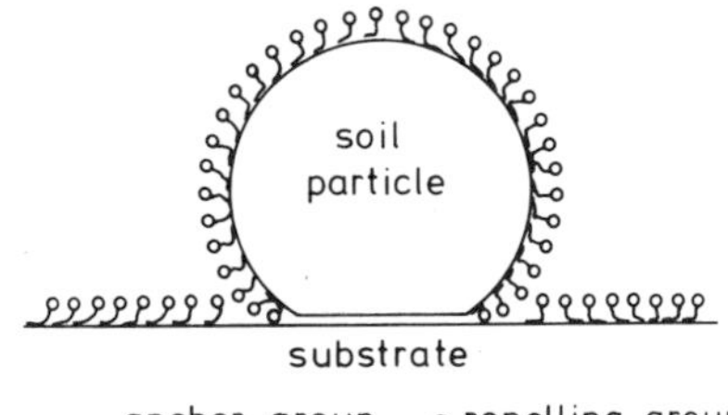

Fig. 2. Adsorption of stabilizer tends to separate soil and substrate.

additional mechanic action is required. Water is the usual medium, but dry cleaning in chlorinated hydrocarbons forms a well known alternative.

Given the great variety of soil (oil, coal, denatured proteins etc., etc.) a very general type of stabilizer is required. Fortunately the highly polar nature of water allows the use of amphipolar detergents, the non polar half acting as anchor group, not on account of a special attraction to soil and fabric, but because it is squeezed out of the water. The stabilizing group may be ionic (classical soaps and its synthetic anionic and cationic variants) or non-ionic, in particular polyethylene oxide.

Whereas the suspensions and emulsions treated in the examples 1-4 are prepared to be used, the dispersions made in washing are intended to be thrown away. An analogous case of "throw away emulsions" is the cleaning up of oil spills by dispersion. Here the use of oil soluble, rather than water soluble surfactants appears to be the correct strategy to avoid loss of active material, and wave and windaction rather than spontaneous emulsification should cause the dispersion of the oil.

The use of "dopes" (= suspension stabilizers) in motoroil [11], in order to keep carbon and other products of incomplete combustion suspended as very fine particles, rather than allowing them to agglomerate and cause wear of engine surfaces is another example of throw away dispersions. Recently, moreover, the suspension of large quantities of coal in oil has obtained a great deal of interest as a way to substitute cheap coal for expensive oil in fuels.

III. APPLICATIONS OF DISPERSIONS

The reasons for using stable dispersions may be quite varied. In milk and other foodstuffs the high interfacial area and the small particle size promote digestibility. In blood the small particles may pass through the capillaries without clogging them.

Moreover oxygen exchange between erythrocytes and plasma requires only very short diffusion pathways. In painting and writing the ease in bringing colored material in a desired pattern to a surface is the essential asset and in drilling muds it is the peculiar rheological behavior of suspensions.

On the other hand, from Perrin's time on, isodisperse systems have been powerful tools in the hands of pure scientists. Perrin himself [12] used redispersed natural latex, or carefully fractionated mastix emulsions to derive Avogadro's constant from Brownian phenomena. Various isodisperse systems were used in conjunction with theoretical work on light scattering [13] and, recently, synthetic latices allowed a critical test of an extension of Smoluchowski's theory of collision frequencies [14].

One of the simplest applications is the *latex dipping process*, in which a smooth, non porous mold, either preheated or covered with a thin layer of a coagulant, is dipped in a concentrated (e.g. 60%) latex. After retraction from the latex bath the adhering layer can be coalesced, dried and vulcanized by the application of moderate heat.

Another simple process is *slip casting* of ceramics. Here a stable aqueous suspension is poured into a porous mold (e.g. plaster of Paris), which takes up part of the water, thus concentrating and rigidifying the suspension, which then acquires sufficient strength for further handling.

In applying a *paint* layer, the suspension or emulsion must be quite fluid during the application by brushing or spraying, but must rigidify soon after, to prevent sagging. The final strength is then obtained by further evaporation of solvent and by polymerization and crosslinking reactions, in which coalescence of oil or latex particles may have an essential role.

The preparation of *magnetic tapes* starting from a suspension of the magnetic particles poses a particularly tough colloid problem since magnetic forces are strong and far reaching, so that a fairly long range and strong repulsion, necessarily based on adsorbed molecules, must help to combat formation of large aggregates, which would spoil the degree of resolution of the signal.

A more sophisticated way of applying layers of small particles, usually, but not exclusively from emulsion or latex paints, is *electrodeposition* [15]. In this most interesting process, highly concentrated layers of stable particles are deposited by d.c. on a metal or metallized surface. The high electrical resistance of a compact layer of particles tends to make the thickness of the layer rather uniform, even in recesses which are not easily reached by spraying or brushing. The electrolysis products, causing a shift in pH or simply an increase in ionic strength, will then promote coagulation of the deposit, the whole process combining the advantages of a dense sediment, which can be only produced from a stable suspension (or emulsion), and good adherance based on coagulation. Of course, the strength of the layer may be further increased by additional heat treatment.

In *drilling muds* [16] a number of properties must be combined. They must be fluid and have a high heat capacity to cool the drilling bit. They must have a high density to withstand the pressure in the formation. This high density and a high rate of flow help to carry the chips loosened by the drilling to the surface. If the drilling is interrupted, as it must be from time to time, the mud must gel in order to prevent accumulation of the chips and the suspended particles at the bottom of the hole. Finally the excess pressure in the mud column over the contents of the formation pushes the mud into the formation. The clay particles in the mud must then plaster out and form a virtually impermeable lining on the wall of the hole. The combination of these divergent properties requires the mud to be a thixotropic suspension with low viscosity when kept in motion but forming rather quickly a weak gel on standing. The addition of some surfactant and a relatively high pH are often essential to obtain the required rheological behavior.

A general remark, to apply to nearly all applications of suspensions, is that the rheology is important, that the systems should certainly not be grossly flocculated, but colloidal stability with a small degree of thixotropy (sometimes called "weak flocculation") is the favored state.

IV. COAGULATION OF DISPERSIONS

As mentioned in the introduction it is often necessary to destroy the dispersed state by inducing coagulation and/or coalescence. One has the classical examples of making cheese, butter or cottage cheese, the clarification of beer, or the coagulation of natural rubber latex.

A very large scale example is found in the demulsification and separation of *natural petroleum emulsions* [9]. A sizable fraction of all crude oil is produced in the form of emulsions of the water in oil type. Before further processing and even before transport the water content has to be brought down below 1%. Although such emulsions as a rule are not very stable, a variety of methods is used to accelerate the separation, among them the addition of chemicals which counteract the natural emulsifiers (e.g. surfactants which - if present alone - would promote oil-in-water emulsions). Passage through water wettable filter beds on which the water droplets collect and coalesce and electrical demulsification are other methods. The electric field acts in this case not so much by causing electrodeposition of the water droplets as by polarizing them so that their mutual attraction causes coalescence.

Another major area in which suspensions are destroyed is that of *coal washeries* [17]. Coal as produced from the mine usually contains clay or claylike material and fine coal dust. If these are separated from the larger lumps of coal by washing with water a nasty black suspension is produced, which should not be discharged into public waters, and which, if left to itself, takes a long time to separate by sedimentation. The application of sensitized flocculation will solve the problem. Sensitized flocculation involves the addition of a small concentration of a polymer, usually a polyelectrolyte which may form bridges because molecules get adsorbed with two anchor groups on two different particles. Agglomerates are formed, big enough to settle rapidly, with enough mutual adherence to include all, also the very small, particles and open enough to allow rapid filtration. The precipitate then contains enough coal to be of value as fuel and the remaining water is clear and not unduely contaminated by chemicals.

This sensitized flocculation [18], discovered long ago (about 1900) as a relative rarity in systems showing protective action at higher polymer concentration, may well become one of the most attractive methods for destroying the stability of suspensions, on account of the small amounts of agent required, and of the openness and good coherence of the flocs. Clarification of sewage and other waste waters may be based upon the same principles.

Soil improvement by partially saponified polyacrylamide is another application of sensitizing action.

Froth flotation [19], the process in which yearly millions of tons of minerals are separated and concentrated, may be regarded as heterocoagulation between fine mineral particles and gas bubbles. Variants of the same technique have been used to remove other particulate materials from suspension and fairly recently the use of small bubbles [20] (diameter about 50 u) has allowed the efficient removal of bacteria and particles of colloidal dimensions. Particles adhere to a water/gas interface when they are hydrophobic. Some kinds of particles are inherently hydrophobic, but more often hydrophobicity is introduced by the adsorption of amphipolar solutes on the particles.

Since this adsorption may be selective, flotation can be used to collect one particular kind of particles from a mixture. Selective flotation of course requires the suspension of particles to be non-flocculated or if it is flocculated, the particles should be fairly large, so that the floccules formed are continuously broken up into individual particles by the agitation. Flotation is a rate process and can only be successful if sufficient time is allowed for the particles to become attached to the gas bubbles before these have reached the foam layer floating on the suspension. There is also a thermodynamic condition [21] which must be satisfied for particle bubble attachment to occur. This is the requirement of a finite contact angle θ.

Fig. 3 shows two particles, A and B, attached to a bubble, particle A being very hydrophobic and having a large contact angle θ_A, whereas particle B is less hydrophobic and has a small contact angle. Particle A will adhere more strongly to the bubble than particle B. Young's equation (9) gives the relation between the contact angles and the respective interfacial tensions (γ_{SL}, γ_{SG} and γ_{LG}):

$$\gamma_{LG} \cos \theta = \gamma_{SG} - \gamma_{SL} \tag{9}$$

If the particle is originally hydrophilic ($\theta = 0$) and the amphipolar solute (the collector)is added, the three interfacial tensions will be modified by adsorption according to the Gibbs adsorption equation,

$$\gamma = \gamma^{o} - \int_{c=0}^{c\ \mathrm{final}} \Gamma \, d\mu_{\mathrm{solute}} \tag{10}$$

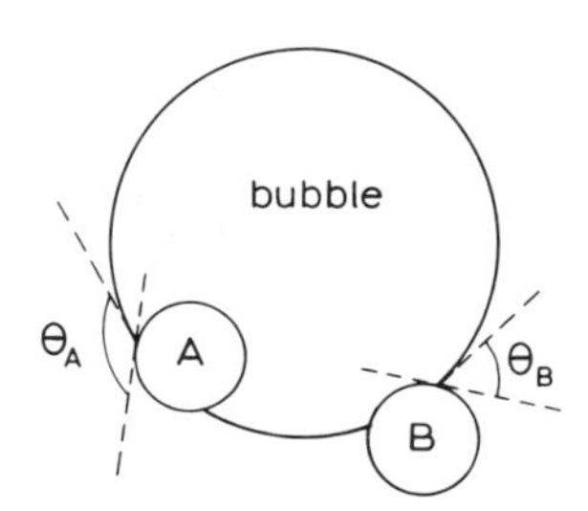

Fig. 3. Particles with large (θ_A) and small (θ_B) contact angles respectively.

An increase of the contact angle θ (a decrease of $\cos\theta$) can only be obtained in case γ_{SG} is decreased more than γ_{SL}, or if:

$$\int \Gamma_{SG}\, d\mu_{solute} > \int \Gamma_{SL}\, d\mu_{solute} \qquad (11)$$

or

$$\int (\Gamma_{SG} - \Gamma_{SL})\, d\mu_{solute} > 0 \qquad (12)$$

Thus, an essential requirement for flotation is that the adsorption of the collector on the solid/gas surface is larger than or occurs at lower concentrations than the adsorption at the solid/liquid interface. The point is stressed because it is sometimes forgotten that the substance causing the hydrophobicity has to reach the solid/gas interface and stay there.

Finally when the material to be floated has accumulated in the foam, the next step in the process must be separation of the foam from the bulk and breaking of the foam.

The above may remind us that the flotation process really applies the whole bag of tricks of surface and colloid science, such as stability vs. flocculation, play with contact angles, breaking of foams and considerations of frequencies of collisions between bubbles and particles.

By discussing a few rather divergent cases of application of colloid science and by showing how often they are based upon the fundamentals of our science, I hope to have strengthened the conviction that ours is a field in which fundamental and applied science often go hand in hand.

REFERENCES

1. H.R. Kruyt, ed. *Colloid Science*, Vol. 1, p. 58 ff. Elsevier, Amsterdam, 1952.

2. C.E. Engel, ed., *Photography for the scientist*, Ac. Press, London, 1968.
C.E.K. Mees and T.H. James, eds., *Theory of the Photographic Process*, 3rd ed. Mac Millan, New York, 1966.
T.G. Bogg, M.J. Harding and D.N. Skinner, *J. Photogr. Sci.*, 24: 81 (1976).
3. F.A. Bovey, I.M. Kolthoff, A.I. Medalia and E.J. Meehan, *Emulsion Polymerization*, Interscience, New York, 1955.
C.P. Roe, *Ind. Eng. Chem.*, 60: 20 (1968).
J.C.H. Hwa and J.W. Vanderhoff, eds., *New Concepts in Emulsion Polymerization*, J. Polymer Sci., C27 (1969).
4. R. Zsigmondy, *Z. physik. Chem.*, 56: 65 (1906).
V.K. La Mer and M.D. Barnes, *J. Colloid Sci.*, 1: 71 (1946).
H. Reis and V.K. La Mer, *J. Chem. Phys.*, 18: 1 (1950).
H.L. Frisch and F.C. Collins, *J. Chem. Phys.*, 20: 1797 (1952); 21: 2158 (1953).
5. M. von Smoluchowski, *Physik. Z.*, 17: 557, 585 (1916); *Z. physik. Chem.*, 92: 129 (1917).
6. J. Halpern, *J. Chem. Education*, 45: 372 (1968).
7. J.W. Vanderhoff, J.F. Vitkuske, E.B. Bradford and T. Alfrey, jr. *J. Polymer Sci.*, 20: 225 (1956).
E.B. Bradford, J.W. Vanderhoff and T. Alfrey, jr., *J. Colloid Sci.*, 11: 135 (1956).
J.W. Vanderhoff, E.B. Bradford, H.L. Tarkowski and B.W. Wilkinson, *J. Polymer Sci.*, 50: 265 (1961).
G.W. Poehlein and J.W. Vanderhoff, *J. Polymer Sci., Polymer Chem. Ed.*, 11: 447 (1973).
8. G.D. Parfitt, ed., *Dispersion of powders in liquids with special reference to pigments*, 2nd ed., Applied Science Publishers, London, 1973.
J.L. Moilliet, B. Collie and W. Black, *Surface Acitivity*, 2nd ed., Spon, London, 1961, Ch. 7.
9. P. Becher, *Emulsions. Theory and Practice*, 2nd ed., A.C.S. Monograph 162, Reinhold, New York, 1965.
P. Sherman, ed., *Emulsion Science*, Ac. Press, New York, 1968.

A.L. Smith, ed., *Theory and Practice of Emulsion Technology*, Ac. Press, New York, 1976.

10. J.L. Moilliet et al., cited under [8], Ch. 8.
11. E.L. Mackor, *J. Colloid Sci.*, 6: 492 (1951).
    E.L. Mackor and J.H. van der Waals, *J. Colloid Sci.*, 7: 535 (1952).
12. J. Perrin, *Les Atomes*, 9me mille, Librairie Félix Alcan, 1920.
13. M. Kerker, *The Scattering of Light and Other Electromagnetic Radiation*, Acad. Press, New York, 1969, Ch. 7.
14. J.W.Th. Lichtenbelt, C. Pathmamanoharan and P.H. Wiersema, *J. Colloid Interface Sci.*, 49: 281 (1974).
15. H. Koelmans, *Philips Res. Reports*, 10: 161 (1955).
16. H. van Olphen, *An introduction to clay colloid chemistry*, Interscience, New York, 1963, pp. 127-129, 139-143, 284.
17. R.A. Henry, *French patent*, 658306 (1928); *Engineering*, 138: 213, 293 (1934); 142: 607 (1936).
    H.A.J. Pieters and J.W.J. Hovers, *Dutch patent*, 61401.
18. H. Freundlich, *Kapillarchemie*, *Band II*, 4th ed., Akad. Verlagsges., Leipzig, 1932, p. 468 ff.
    V.K. La Mer and R. Smellie, *J. Colloid Sci.*, 11: 704 (1956); 13: 589 (1958).
19. A.M. Gaudin, *Flotation*, 2nd ed., Mc. Graw Hill, New York, 1957.
20. E.A. Cassel, K.M. Kaufman and E. Matijević, *Water Research*, 9: 1017 (1975).
    J.B. Melville and E. Matijević, *J. Colloid Interf. Sci.*, 57: 94 (1976).
21. P.L. de Bruyn, J.Th.G. Overbeek and R. Schuhmann, Jr., *Transactions AIME, Mining Engineering*, 1954, p. 519.

LATICES

ELECTROPHORETIC DEPOSITION OF LATEX ON STEEL

Dennis C. Prieve and John P. Carrera*

Department of Chemical Engineering
Carnegie-Mellon University
Pittsburgh, Pennsylvania

I. ABSTRACT

Up to 60 vdc is applied across an aqueous coating bath between a 4 cm-diameter, rotating, steel-disc anode and a stationary steel-disc cathode, separated by 1-10 cm. The coating bath contained 0.1-1.0 wgt % of commercial 65/35 styrene-butadiene latex, $HC\ell O_4$ to give pH 2-7, 2 ml of 3.5% H_2O_2, and distilled water to give 1 Kg total mass. Exposure times of 1-15 min and rotation speeds of 0-600 rpm produce films with thickness of up to 0.2 mm when dried. Both the mass of iron dissolved and the mass of polymer deposited on the anode are independently measured together with the electrical current. Results indicate that the mass flux of polymer depositing is proportional to the product of the mass flux of dissolving iron and the mass density of polymer in the bath. The proportionality constant is found to be 2200 cm^3/g. In turn, the rate of iron dissolution is controlled by the conductivity of the coating bath and the applied electric field, and thus is independent of the thickness of the deposited film.

*Currently with Exxon Research & Engineering Co., Baytown, Texas 77520.

II. INTRODUCTION

Electrodeposition from aqueous baths containing dilute solubilized polymer is widely used today to apply a protective primer coating on motor vehicle chassis. Sheppard and Eberlin(1) first investigated electrodeposition of natural rubber latex as early as 1923. A discussion of the overall mechanism is given by Sheppard(2) and Beal(3). Later, Fink and Feinleib(4) systematically explored the electrodeposition of synthetic resins. During the past decade the paint industry has developed the process of electrodepositing water soluble resins. A recent review of this technology is given by Beck(5).

Our interest in this problem stemmed from a recent patent by Steinbrecher and Hall(6) who demonstrated that rates of latex deposition on steel panels from dilute aqueous dispersions could be as large without the applied electric field as has been previously reported with an electric field. If so, what determines the rate of deposition? Conventional mechanisms(2-5) for electrodeposition of polymer from aqueous dispersions or solutions usually involve a step in which ions, generated by electrode reactions, associate with the polymer near the electrode to neutralize the polymer's stabilizing charge and permit deposition. With oxidizable metal electrodes, such as iron or mild steel, application of a negative voltage causes metal ions to be produced by dissolution, whereas with inert electrodes, such as platinum, H^+ or OH^- are produced by electrolysis. Both Fe^{+3} and H^+ have been observed to cause coagulation of sols in much lower concentrations than required by DLVO theory, strongly suggesting their specific adsorption on the surface of the sol particles. Since the solutions used by Steinbrecher and Hall have a much lower pH (usually 3-4) than conventional electrodeposition baths, the acid may induce dissolution of the iron to produce the ions needed to neutralize the polymer's charge.

The present paper is a preliminary report on an experimental and theoretical investigation of the role and relative importance of various mechanisms affecting the transport of iron ions and latex particles during the deposition process, including convection, diffusion, and electrophoretic migration. Unlike most previous investigations, a rotating-disc electrode is used so that both fluid convection and the electric field can be mathematically described. Under the conditions of the present experiment, the iron dissolution rate is controlled by the conductivity of the bath whereas the latex deposition rate is proportional to the product of dissolution rate and the bulk latex concentration.

III. EXPERIMENTAL

A. Apparatus

The metal workpiece that is to be coated is machined in the shape of a circular disc which can be mounted in the recess provided in the base of an epoxy cone, as shown in Fig. 1. A small flat-head screw (not shown), driven through the center of the disc, holds the disc in place and assures good electrical contact with

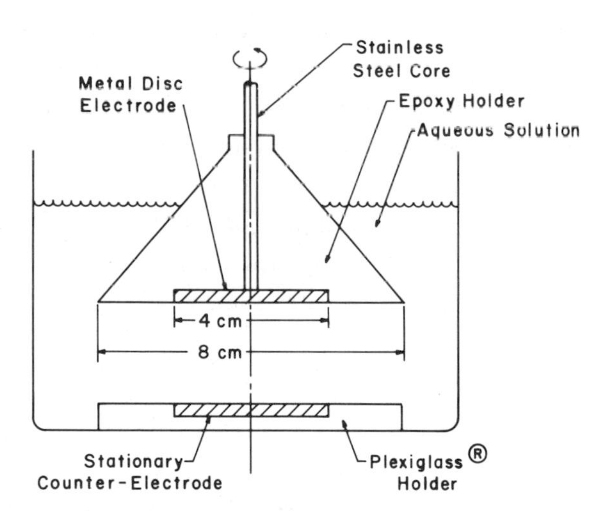

Fig. 1. Schematic drawing showing construction of rotating-disc electrode. Voltage is applied between rotating and stationary disc electrodes.

the stainless steel core. The entire cone assembly is rotated by a Pine Instrument Co. Model CDR rotator, which is driven by a variable-speed motor. With the rotating cone design shown in Fig. 1, hydrodynamic edge effects should not interfere with the processes occurring on the active metal surface.

An electrical potential of up to 60 volts is applied between the rotating metal disc electrode and the stationary counter-electrode using a Hewlett-Packard Model 3465B d.c. power supply. In all the experiments reported below, the rotating disc is the anode with the power supply operating in the constant-voltage mode. Current generally decreased slowly and at a decaying rate during the first minute of operation, although the total variation over any run is less than a few percent of the initial value. This is probably due to polarization within the porous polymer film accumulating on the anode. Total charge passed is determined by recording the current on a strip-chart recorder, then graphically integrating over the total exposure time.

B. Pretreatment of Metal Surfaces

Metal discs, on which coatings were applied, were sliced from a single rod of mild steel, then individually machined to fit smugly into the base of the epoxy cone. The exposed face of each disc was mechanically polished to a smoothness estimated at 0.8μm. After mounting, the metal surface was cleaned with a detergent solution, then rinsed with acetone and dried. Care was taken to incline the base of the cone during immersion in the coating bath so that no gas bubbles were trapped. Each disc was used only once.

C. Preparation of the Coating Bath

Due to the large amounts of latex consumed in each run and the expense of well-characterized latex, a commercial latex was arbitrarily selected -- Goodyear's product LPR4121A. This Pliolite resin latex is a copolymer containing 65% styrene and 35% butadiene, stabilized by anionic surfactant. As received, it has pH 9.2, specific gravity of 1.01, and is 49.8% solids. A 10% stock solution was made by dilution with distilled water.

For each run, 1 Kg of coating bath was prepared by weighing the appropriate amount of stock solution into a beaker, adding 2 ml of 3.5% H_2O_2, then adding distilled water to a final total weight of 1 Kg. The bath is then titrated to the desired pH (usually 3.5) with $HC\ell O_4$. Peroxide was added to suppress H_2 gas formation at the cathode, whereas perchloric acid is used because the perchlorate ion does not seem to form ion complexes with ferrous or ferric ions.

D. Procedure

After separately bringing the cone assembly and coating bath to the desired operating temperature (usually 30°C) using a constant-temperature water bath, the cone is immersed in the coating bath, attached to the rotator and the electrode separation distance is set (usually 1.75 in). Then rotation speed is adjusted (usually to 200 rpm) and the voltage is applied for the desired exposure time (usually 5 min). After exposure, the cone assembly is disconnected from the rotator and removed from the coating bath. Orienting the base of the cone parallel to a vertical plane, the outer circumference of the cone is touched to an adsorbent paper towel to remove drops of coating bath clinging to the epoxy. The polymer film adhering to the steel disc is separated from the disc using a directed jet of distilled water from a squeeze bottle and placed in a glass dish to be dried in an oven overnight and weighed on an analytical balance. The steel disc is removed from the epoxy cone, dried and weighed. Thus the mass of polymer deposited and the mass of iron dissolved were independently measured.

IV. EXPERIMENTAL RESULTS

Figure 2 shows the mass of iron dissolved and the mass of coating deposited as a function of exposure time. Both increase

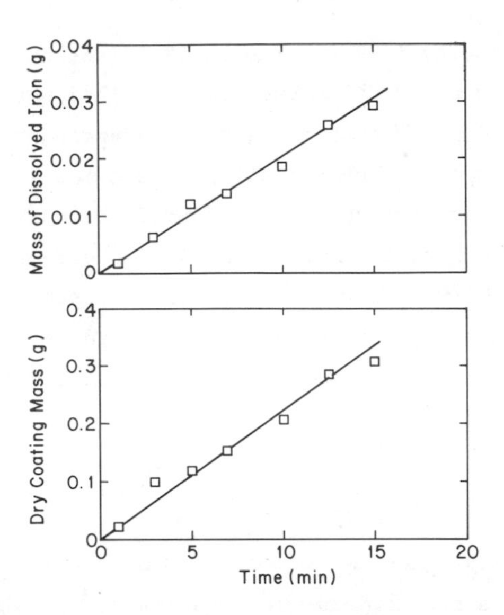

Fig. 2. Variation in mass of iron dissolved and mass of coating deposited as a function of exposure time. Linear growth of coating mass rules out transport of iron ions through film as the rate-limiting step. Conditions of experiment: 30 vdc, 0.5 wgt % latex, pH 3.5, 200 rpm, 30°C, and 1-7/8 inch spacing of electrodes.

linearly with time, indicating that the rates are independent of film thickness. This suggests that transport of iron ions through the depositing film is not the rate-determining step, although the film grows to a thickness of 0.5 mm (estimated using a porosity of 0.5 when wet).

The effect of applied voltage at two different electrode separation distances is given in Fig. 3. At both separations the masses seem to be proportional to the applied voltage. Data for the smaller separation show considerably more scatter about a straight line, which is probably due to the difficulty of reproducing the electrode spacing. Linearity of this data would be expected if electrophoretic migration is the dominant mechanism. However, then the rates would be proportional to the ratio of voltage to separation distance (the electric field) and all of the data would fall on the same straight line. Other experimental

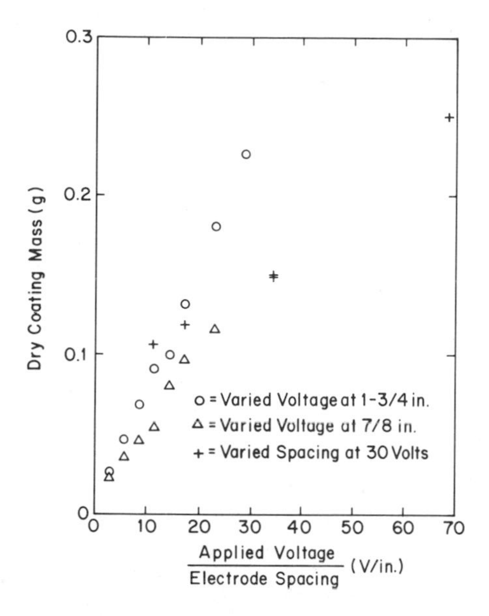

Fig. 3. Variation in mass of iron dissolved and mass of coating deposited as a function of applied voltage and of the electrode spacing. Other experimental conditions are the same as Fig. 2 with an exposure of 5 min.

factors which might account for absence of good correlation with the electric field include the difficulty in aligning the electrodes parallel to each other and the nonequal surface area of the two electrodes. Both factors could cause distortion of electric field lines which is expected to be largest at small separations.

During the initial experiments (not reported here), the rates were found to be reproducible if repeated on the same day, but not if measurements were taken several days apart. This was attributed to variations in room temperature and led to the use of a constant-temperature water bath in all subsequent runs. The effect of temperature is shown in Fig. 4. An increase of 10°C enhances the rates by 20-30%, enough to explain day-to-day variations in the absence of temperature control. Such enhancement is probably due to changes in ion and particle mobilities with temperature rather than changes in reaction rates, since the latter can easily double with an increase of 10°C.

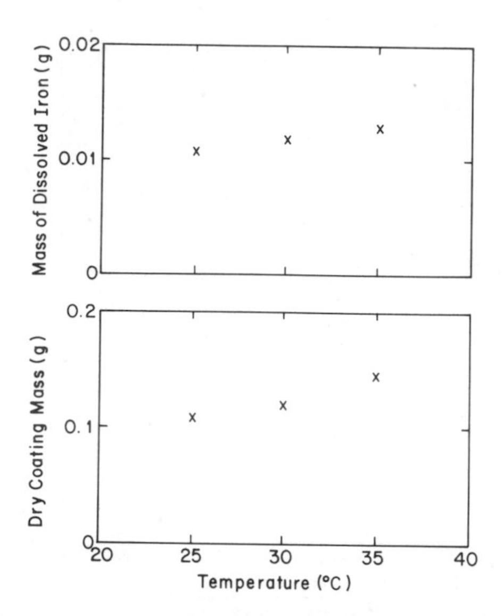

Fig. 4. Variation in mass of iron dissolved and mass of coating deposited as a function of temperature. Other experimental conditions are the same as Fig. 2 with an exposure of 5 min.

Figure 5 indicates the effect of disc rotation speed. The iron dissolution rate is practically independent of speed for more than 100 rpm -- further indicating that electrophoretic migration, rather than convection, is the dominant transport mechanism. Low dissolution rates in the absence of forced convection are probably due to polarization near the electrode. On the other hand, the latex deposition rate is highest at 0 rpm and decreases slowly to about half the maximum rate at 200 rpm. Films deposited at low speeds are loose and spongy, suggesting a significantly higher water content. Considerable difficulty in reproducing the deposition rate was experienced near 500 rpm. This speed corresponds roughly to the critical disc Reynolds number at which the laminar/ turbulent transition occurs. Decreases in deposition rate with rotation speed rule out the particle's diffusion boundary-layer resistance as the rate-limiting step.

Lowering the pH caused substantial increases in dissolution and deposition rates, as shown in Fig. 6. However, at pH's of 3.5

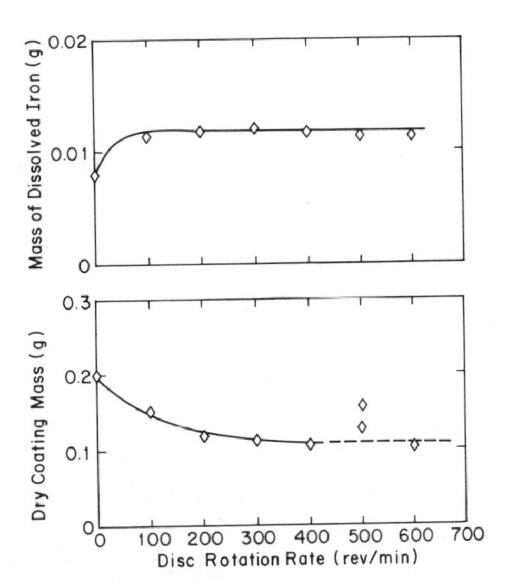

Fig. 5. Variation in mass of iron dissolved and mass of coating deposited as a function of disc rotation speed. Other experimental conditions are the same as Fig. 2 with an exposure of 5 min.

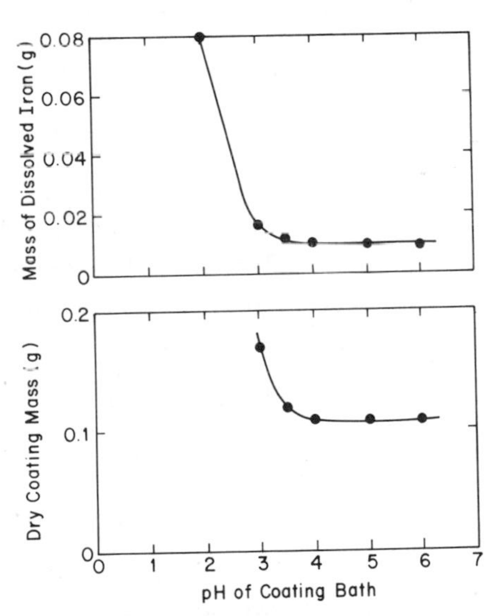

Fig. 6. Variation in mass of iron dissolved and mass of coating deposited as a function of initial pH of the coating bath. After exposure the pH is higher. Other experimental conditions are the same as Fig. 2 with an exposure of 5 min.

or larger, the rates are independent of pH, suggesting that the concentration of H^+ is important only when it is very large. One explanation is that the dissolution rate is determined by the conductivity of the bulk of the coating bath. In other words, the rate of iron dissolution is limited by how fast charge can be transported through the bulk of the solution between the electrodes, rather than by the electrode reactions. The increase in dissolution rate (by nearly a factor of 5), coinciding with a lowering of the pH from 3 to 2, may simply be explained as an increase in the concentration of charge carriers (H^+) in the bath.

No coating weight is reported at pH 2 because the latex in the coating bath coagulated shortly after the voltage was applied. Since the latex is normally stable for long periods at pH 2, and since the iron dissolution is much higher in this case, the coagulation is probably caused by adsorption of hydrolyzed iron ions, which could neutralize the surface charge of the latex particles. Matijević(7) and O'Melia and Stumm(8) have explained destabilization of several types of sols by adsorption of hydrolyzable metal cations, including iron.

Finally the concentration of latex was varied. Since the stock solution is basic, more moles of acid have to be added to lower the pH to the same final value for higher concentrations of latex. Then the number of charge carriers in the bath would also increase. To avoid changing both latex concentration and conductivity simultaneously, the procedure is slightly modified. After titrating to a pH of 3.5, a solution of $NaClO_4$ is added drop by drop until the conductivity (indicated by the electrical current) is the same as for the solution having the highest latex concentration. Figure 7 summarizes the result. As expected, the iron dissolution rate is independent of latex concentration. On the other hand, the rate of coating deposition is proportional to latex concentration. This first-order concentration dependence suggests that particles of latex deposit individually, rather than as clusters of several particles, since rates of formation of the latter would be proportional to the square of concentration.

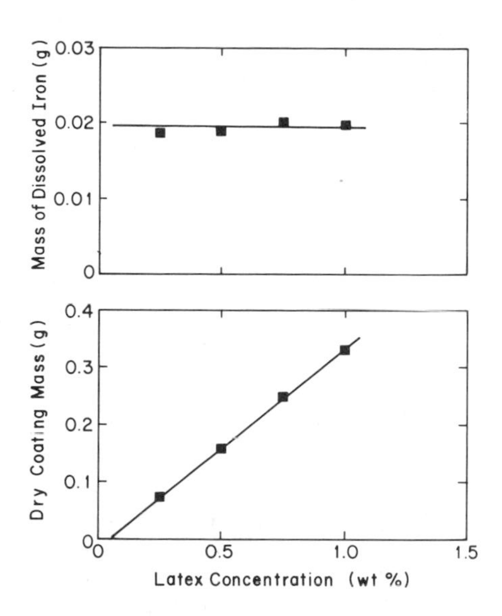

Fig. 7. Variation in mass of iron dissolved and mass of coating deposited as a function of latex concentration in coating bath. Other experimental conditions are the same as Fig. 2 with an exposure of 5 min.

V. DISCUSSION

Collectively, Figs. 2-7 seem to indicate that the rate of latex deposition is proportional to the latex concentration, with the proportionality constant, in turn, being proportional to the iron dissolution rate. Figure 8 cross-correlates the mass of polymer deposited with the mass of iron dissolved. Despite the variety of parameters altered, nearly all the data fall on a straight line, suggesting that the rate of deposition is proportional to the rate of iron dissolution. The slope of the line drawn in Fig. 8 corresponds to a yield of 11g of dry polymer deposited per gram of iron dissolved. Points deviating appreciably from this line correspond either to latex concentrations other than the usual 0.5% or to zero rotation speed. Dividing all the coating masses by the corresponding latex concentration would bring all the points into reasonable agreement with the straight line, except for the singular point corresponding to no stirring. In the latter case, the

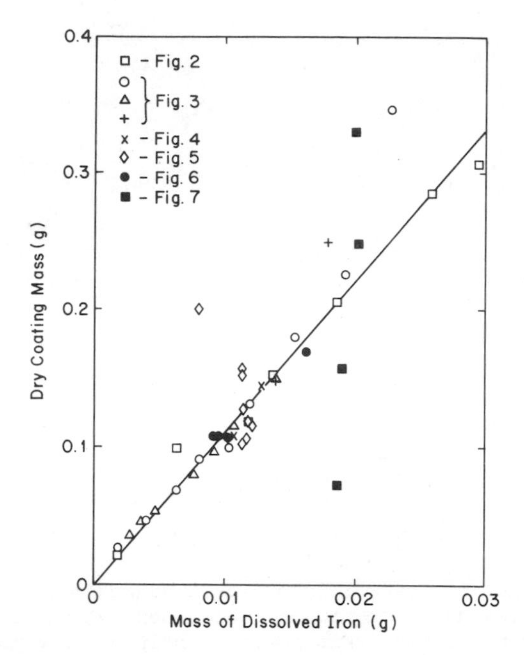

Fig. 8. Cross-correlation of the mass of coating deposited with the mass of iron dissolved, using all the data of Figs. 2-7.

rate-controlling step is apparently different. Otherwise the rate of deposition is simply:

$$n_p = - K\, n_{Fe}\, \rho_p \qquad [1]$$

where n_p and n_{Fe} are the mass fluxes in the +z-direction of polymer and dissolved iron, ρ_p is the mass density of polymer in the bath, and $K = 2200\ cm^3/g$.

In turn, the iron dissolution rate, n_{Fe}, seems to be controlled by the rate at which charge can be conducted through the well-stirred bath. This should be controlled by the concentration, charge, and electrophoretic mobility of charge carriers in the bath. Total charge passed and the mass of iron dissolved were independently measured in all of the experiments. A cross-correlation of these two quantities is shown in Fig. 9. These data indicate a linear relationship. The straight line drawn through the

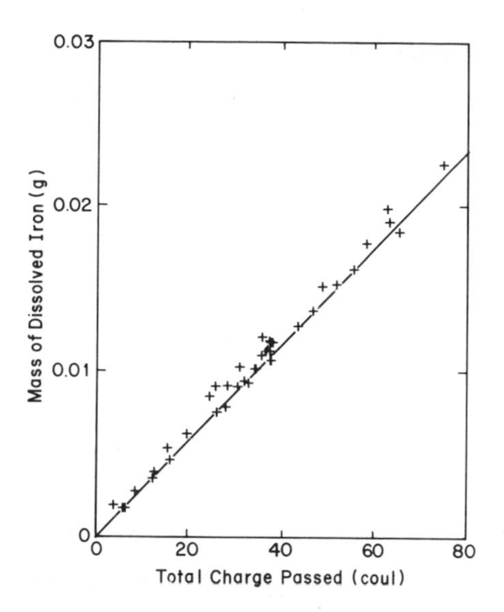

Fig. 9 Cross-correlation of the mass of iron dissolved with the total charge passed through the cell, using all the data of Figs. 2-7. Slope of straight line corresponds to 2 equivalents per mole of iron.

data has a slope corresponding to 2 equivalents per mole of iron, which indicates that the only anodic reaction is

$$Fe(s) \rightarrow Fe^{+2}(aq) + 2e^{-} \qquad [2]$$

In the presence of H_2O_2, ferrous ions will be further oxidized to ferric, although the slope of the line in Fig. 9 indicates that this does not occur on the anode.

Since the ferrous or ferric ions do not penetrate far into solution outside the deposited film before they are adsorbed on latex particles, other charged species such as the latex particles or other ions must carry the current across the bath to the other electrode. Discharge at the cathode probably occurs by:

$$2H^{+}(aq) + H_2O_2(aq) + 2e^{-} \rightarrow 2H_2O(\ell) \qquad [3]$$

in the presence of peroxide. Because hydrogen ions are being consumed at the cathode, the pH can be expected to increase with time. Indeed, the pH of the coating bath measured after exposure was as much as 0.5 higher than before exposure. Using the stoichiometry of reactions [2] and [3], the amount of H_2O_2 added is sufficient to oxidize 0.12g of iron to Fe^{+2} or 0.08g to Fe^{+3}. In the absence of H_2O_2, reaction [3] is probably replaced by:

$$2H^+(aq) + 2e^- \rightarrow H_2(g) \qquad [4]$$

No gas evolution was observed, except in a test case in which H_2O_2 was omitted. A summary of the electrolytic circuit described above is depicted in Fig. 10.

Of crucial importance are the microscopic events which occur at the advancing front of the deposited polymer film. At least part of the charge of the latex particles must be neutralized before the particles may come together to form a film. Adsorption of species like $FeOH^{+2}$ and $Fe(OH)_2^{+1}$ on the particle surface is probably responsible for charge neutralization. A mathematical analysis of the process should consider electroconvective diffusion of both iron ions and latex particles. Adsorption of iron ions on latex outside the film requires consideration of a homogeneous reaction which consumes iron when writing the equations to

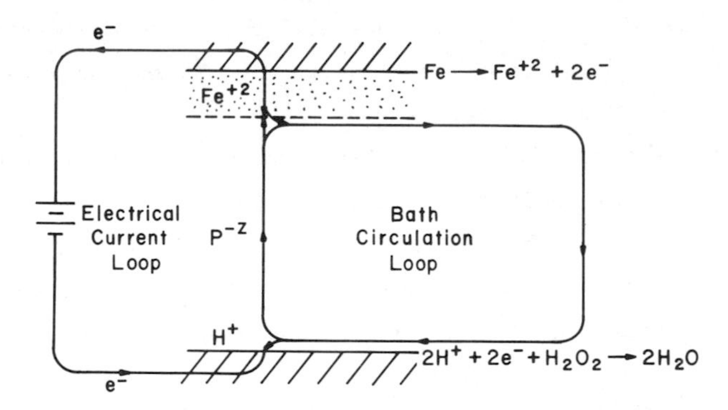

Fig. 10. Schematic circuit diagram for cell, showing the probable charge carrying species. Arrows point in direction of mass transfer (not charge transfer).

model the transport process, as well as a variation in the electrophoretic mobility of the latex particles. This reaction will result in coupling of the two differential equations for the two transporting species. Such an analysis is currently in progress.

Besides gathering more data of the type described here, additional experiments are needed to complete the analysis. Independent measurement of the electrophoretic velocity and size of the latex particles, measurement of the adsorption isotherms of hydrolyzed iron cations on the polymer's surface, as well as studies of the stability of these latices in the presence of iron cations are currently in progress or planned.

A variety of limiting behavior is likely to be revealed in the general mathematical analysis which, in turn, will suggest new experiments. The product will be an improved understanding of electrode processes involving latex particles, which should suggest new methods for applying polymer films on conducting surfaces, provide a new tool for studying hydrosol adsorption, and perhaps yield a new technique for measuring electrophoretic mobilities.

VI. ACKNOWLEDGEMENTS

Part of this work was supported by the Materials Research Laboratory Section, National Science Foundation under Grant No. DMR72-03297 A03.

VII. REFERENCES

1. Sheppard, S.E. and L.W. Eberlin, U.S. Patent 1476374 (1923).
2. Sheppard, S.E., Trans. Am. Electrochem. Soc. 52, 47 (1927).
3. Beal, C.L., Ind. Eng. Chem. 25, 609 (1933).
4. Fink, C.G. and M. Feinleib, J. Electrochem. Soc. 94, 309 (1948).
5. Beck, F., Prog. in Organic Coatings 4, 1 (1976).

6. Steinbrecher, L. and W.S. Hall, U.S. Patent 3585084 (1971).
7. Matijević, E., J. Colloid Interface Sci. 58, 374 (1977).
8. O'Melia, C.R. and W. Stumm, J. Colloid Interface Sci. 23, 437 (1967).

DISCUSSION

J.Th.G. Overbeek (University of Utrecht, Netherlands)

You seemed to imply that there is a contradiction between the classical explanation of the regular coverage of an irregular surface (covered areas have a higher resistance and push the streamlines away) and the fact that in your own experiments the rate of deposition is nearly independent of time. However, your geometry and field strength are such that the electrophoretic transport towards the electrode is nearly constant since the current is nearly independent of time. I don't see the contradiction. Koelmans (Philips' Research Reports, about 1955) reported electrodeposition data from suspensions in methanol which might be worthwhile to compare with your own data.

Author I did not intend to imply any contradiction. To obtain a uniform coating on an irregular surface, the rate of the processes must be controlled by the resistance of the deposited film. This is the classical explanation which I accept. In our experiments, neither the electrical current nor the latex deposition rate depend on the thickness of the polymer film already deposited. Therefore, under these conditions the rate is not controlled by the film resistance and nonuniform coverage may be expected over irregular surfaces. Your citation of Koelman's paper is appreciated.

A. K. Chatterjee (General Tire & Rubber)

In one slide you showed that polymer deposition was linear with the applied electrical field, but the slope of the lines was dependent upon the electrode configuration. Any rationalization?

Author Iron dissolution and polymer deposition rates were found to be proportional to the voltage applied between the electrodes. While the rates were linearly related to the reciprocal of the electrode spacing at constant voltage, the relationship is not a direct proportionality, as one would expect if the electric field is the important parameter. Any effect which results in non-vertical field lines could be responsible (such as non-parallel electrodes or unequal electrode areas).

Note added by authors after meeting. Further experiments were conducted in which greater care was taken in electrode alignment and area matching. The current through simple electrolyte solutions (no latex) was again measured as a function of applied voltage and electrode spacing. The results were qualitatively unchanged: current was proportional to voltage and linearly related to reciprocal spacing, but was not proportional to reciprocal spacing. We now believe that, when the electrode spacing is comparable to or greater than the electrode diameter, a disproportionate number of field lines emanate from the edge of the disc which are nonlinear. The effect is to make the apparent cross-sectional area of the solution available for conducting current greater than the electrode spacing. Only if the conducting area is constant will a direct proportionality be observed between current and reciprocal spacing. To minimize these electrical edge effects, future experiments will be done with smaller electrode spacing.

John Porst (Pfizer)

Is the latex deposition rate a function of the hydrogen peroxide concentration?

Author All the experiments reported are for a stoichiometric excess of H_2O_2. In a few experiments performed with different peroxide concentrations, approximately the same rate was obtained. However, a systematic variation over a wide range of peroxide concentration has yet to be performed.

A. H. Herz (Eastman Kodak Co)

It would be relevant to know if any latex deposition occurred (a) at the cathode and (b) at the anode where this was a noble metal.

Author No polymer deposition was ever observed at the cathode, presumably because no ferric ions are locally present to neutralize the polymer's negative charge. This negative charge causes the polymer to be repelled by the cathode. No experiments were performed using noble electrodes. However, certain latexes could be expected to deposit on such anodes. For example, at a Pt anode, electrolysis would produce H^+ and O_2 gas. If the latex sol is one which is unstable in acidic solutions, it may deposit on the Pt anode. However, in the presence of evolving gas bubbles, the film so produced may have a large number of defects.

CONTROL OF SURFACE CHARGE ON POLYMER LATEX PARTICLES

Li-Jen Liu and Irvin M. Krieger

Departments of Macromolecular Science and Chemistry
Case Western Reserve University
Cleveland, Ohio

I. INTRODUCTION

In a polymer latex system, as in many other colloidal systems, the principal interactions between the polymer particles can be attributed to Coulombic and Van der Waals forces [1]. Coulombic forces are due to net electric charges of both particles, and decrease with the square of the distance. Van der Waals forces, although also based on electric interaction, are present even between neutral particles. They are always attractive, and decrease more rapidly with distance. Van der Waals forces are due to the interactions of dipoles, which may be either permanent dipoles of polar particles or induced dipoles of nonpolar but polarizable ones.

A latex particle with its double layer is electrically neutral, and therefore exerts no net Coulombic force upon a similar particle situated at a sufficiently large distance. As the particles approach, however, the double layers interpenetrate and rearrange, leading to a net repulsive force. The higher the charge density on the particle surface, the stronger the resulting repulsion between the particles. Therefore the surface charges on the particles play a determining role in the stability of the colloidal system.

It is known that the thickness of the double layer increases with the surface charge density of the latex particle [2,3]. At similar particle and electrolyte concentrations, therefore, the effective volume of the latex particle increases with increasing surface charge density. As a result, a latex with high surface charge density will be more viscous than a similar latex with lower charge. In this way, the surface charge density on the latex particle plays a determining role in the rheological properties of the latex.

Various methods have been tried to control surface charges on polymer latices. In general, one can vary latex surface charge by varying the latex preparative recipe or one can modify the surface charges after the latex has been made. The preparative route is easier, but modification has an advantage if the goal is to produce a series of latices with the same size but different charges.

Smith [4] suggested a way in 1948 to increase surface charge by adding initiator to the reaction medium after polymerization has begun, optimally after the number of particles in the system has become constant. Van den Hul and Vanderhoff [5] found that, as the pH of the polymerization medium is decreased, the proportion of hydroxyl endgroups increases and that of sulfate endgroups decreases. No carboxyl groups were detected in ion-exchanged latices prepared in a nitrogen atmosphere, regardless of the nature of the emulsifier. Stone-Masui and Watillon [6] have prepared polystyrene latices with only strong acid surface charges, as well as latices with both strong and weak acid charges. They found that an increase of the hydrophobic part of the soap leads to higher surface charges. Their data also show that latices prepared by emulsifier-free recipes have relatively higher surface charges. Wu [7] has recently prepared monodisperse polystyrene latices with or without surface carboxyl groups by varying the ingredients charged at the beginning of the reaction for emulsifier-containing and emulsifier-free systems. He obtained carboxyl-free latices from both emulsifier-containing potassium persulfate-sodium bicarbonate systems and ammonium persulfate-ammonium hydroxide-methanol systems.

Fitch and McCarvill [8] synthesized polystyrene latices with only the non-hydrolyzable sulfonic acid groups, by using a bisulfite/Fe(III) redox initiator and sodium dodecyl sulfate emulsifier. They also synthesized polystyrene latices with both hydroxyl and sulfonic acid groups on the latex particle surface using bisulfite/persulfate initiator and sodiu dodecyl sulfate emulsifier, followed by exhaustive hydrolysis of the surface sulfate groups. McCracken and Datyner [9] made monodisperse polystyrene latices by emulsion polymerization of styrene in a solution of methanol and water. They found that the surface charge densities of these latices were higher than in latices prepared by conventional emulsion polymerization.

Juang and Krieger [10] increased the strong acid surface charges of polystyrene latices by incorporation of small amounts of ionic comonomer into their emulsifier-free polymerization systems. Weak acid groups have also been introduced onto latex particles by copolymerization of monomer with an acid group like itaconic acid [11] or methacrylic acid [12].

Chan and Goring [13] sulfonated polystyrene latex in dichloroethane with SO_3/triethylphosphate complex, and were able to increase surface charge with very little change in size and shape. A new method has been described to introduce sulfonic groups in methyl methacrylate-glycidyl methacrylate copolymer [14]. The polymer particles were suspended in water and reacted in a nitrogen atmosphere with sodium sulfite in the presence of a cationic surfactant such as tetra-n-butylammonium bisulfite. The addition of sulfite ions to the epoxy group in the copolymer results in primary sulfonic acid.

The present report includes work with polymer latices synthesized by both emulsifier-containing and emulsifier-free systems. In an emulsifier-containing system, both chemically bound surface charges from initiator and adsorbed surfactant on the latex particle surface help to stabilize the latex. Since there are no surfactant molecules adsorbed on the particle in emulsifier-free latices, chemically bound surface charges determine their stability. We have attempted to control surface charge of our latices by five methods:

(1) use of sulfonated azo-type initiators, namely azo-bis-(sodium 2-methylbutyronitrile sulfonate) (L1] and azo-bis-(sodium isobutyronitrile sulfonate) [L2].

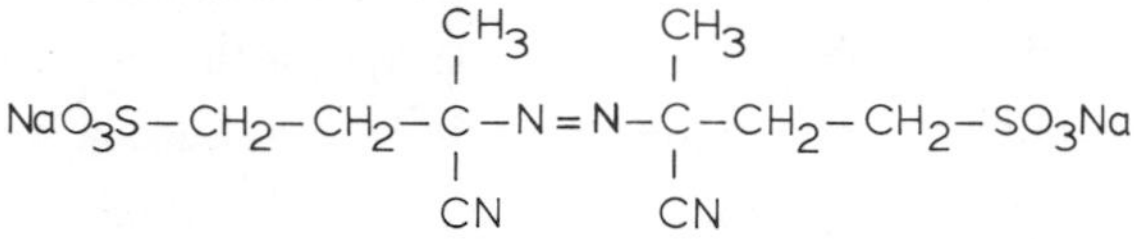

$$NaO_3S-CH_2-CH_2-\underset{CN}{\overset{CH_3}{C}}-N=N-\underset{CN}{\overset{CH_3}{C}}-CH_2-CH_2-SO_3Na$$

[L1]

$$NaO_3S-CH_2-\underset{CN}{\overset{CH_3}{C}}-N=N-\underset{CN}{\overset{CH_3}{C}}-CH_2-SO_3Na$$

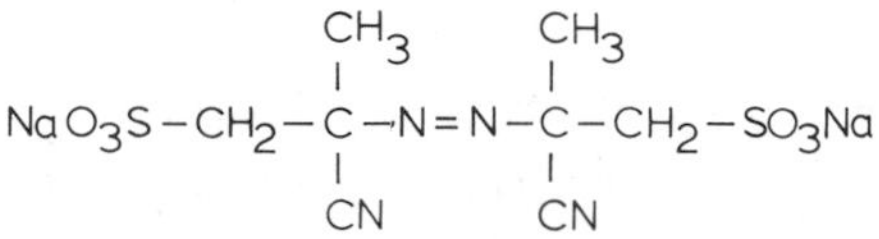

[L2]

(2) use of azo-bis(isobutyramidine hydrochloric acid) [AIBA· 2HCl] as initiator to synthesize latices with positive surface charges.

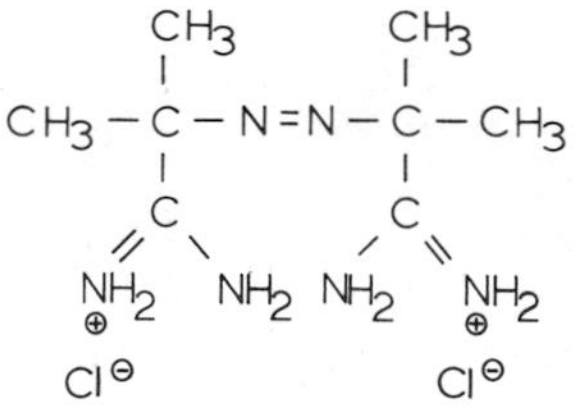

$$CH_3-\overset{CH_3}{\underset{\overset{|}{C}(=\overset{\oplus}{N}H_2\,Cl^{\ominus})NH_2}{C}}-N=N-\overset{CH_3}{\underset{\overset{|}{C}(NH_2)=\overset{\oplus}{N}H_2\,Cl^{\ominus}}{C}}-CH_3$$

[AIBA · 2HCl]

(3) use of ionic comonomers, 1,2-dimethyl 5-vinylpyridinium methyl sulfate [L3], 1-ethyl 2-methyl 5-vinylpyridinium bromide [L4] and sodium vinylbenzyl sulfonate [L5], to synthesize latices with positive or negative surface charges.

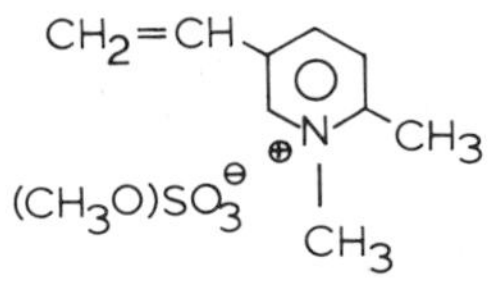

[L3]

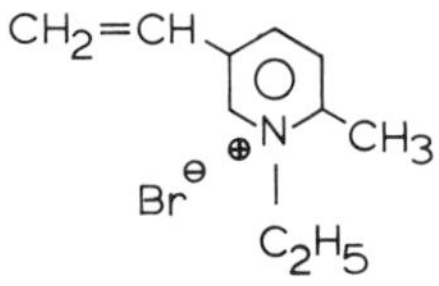

[L4]

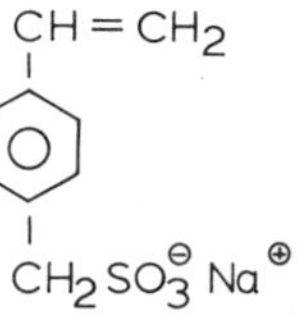

[L5]

(4) addition of charged groups at unsaturations by treating butadiene-containing latex with bisulfite ion in aqueous phase.

(5) conversion of surface chloride into sulfonate by treating poly(vinylbenzyl chloride) latex with sodium sulfite in aqueous phase.

II. EXPERIMENTAL

A. Apparatus

1. POLYMERIZATION REACTORS

Resin Kettle Reactor: The polymerization was carried out in a 2 or 4-liter kettle. The kettle is equipped with a stirrer, a condenser, a thermoregulator, a cooling finger, and a nitrogen gas inlet, as shown in Figure 1. The reaction mixture was purged with N_2 to remove dissolved oxygen and to keep the reaction medium from contacting air. The thermoregulator and cooling finger are used to reduce the temperature during exotherms. The resin kettle was immersed in a constant temperature water bath for the time required for the reaction.

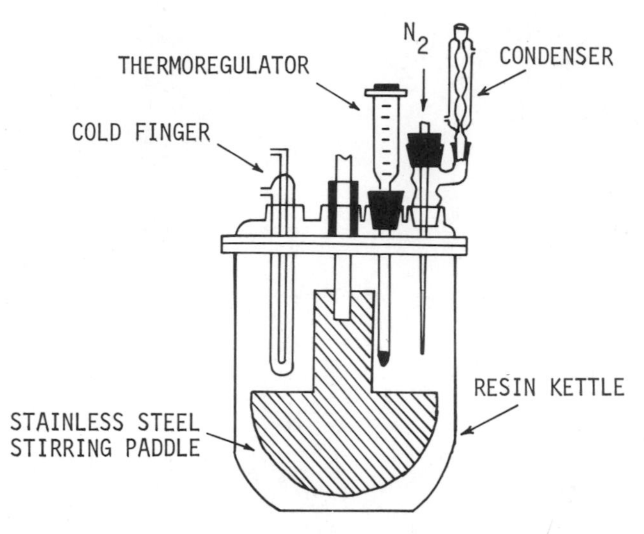

FIG. 1 Resin Kettle Polymerization reactor.

Bottle Reactor: Latices were also produced by bottle polymerization. Twelve-oz citrate of magnesia bottles were charged with ingredients, capped, and then placed in a thermostatted water bath, commonly known as a bottle polymerizer. The polymerizer was equipped with a rotor to which were attached cages for the bottles; as many as 24 bottles could be run simultaneously. The polymerization proceeds in a closed system, with agitation resulting from end-over-end rotation of the bottles.

Three-neck Round-bottom Flask Reactor: 500, 1000, and 2000 ml three-neck round bottom flasks were used to synthesize some of our latices. The flask, equipped with a stirrer (half-moon Teflon blade), a condenser, and a nitrogen gas inlet, is suspended in a constant temperature bath for the time required for the reaction.

2. ELECTRON MICROSCOPY

The electron microscope used was a JEM-100B (Japan Electron Optics Laboratory, Ltd.). The technique used in this study is described in a previous paper [10].

3. CONDUCTIVITY APPARATUS

Conductivity was measured by means of a dip-type glass conductivity cell (Matheson Scientific Co.) and a 60-cycle Wheatstone bridge (Leeds & Northrup Co.) at constant temperature and under nitrogen atmosphere.

4. PH APPARATUS

pH measurements were carried out at 25 ± 0.02°C under nitrogen atmosphere using a Fisher Accument model 520 Digital pH-meter with a Sargent-Welch Miniature Combination Electrode.

B. Reagents

Styrene (reagent, Eastman Kodak Co.) and divinylbenzene (technical, Dow Chemical Co.) were passed through an activated alumina column before use to remove inhibitor. Potassium persulfate was analytical grade (Mallinckrodt Chemical Works), used as received. AIBA•2HCl was kindly supplied by Eastman Kodak Co. and used as received. Initiator L1 was synthesized from ethyl acetoacetate (reagent, Aldrich Chemical Co., Inc.), sodium formaldehyde bisulfite (practical, Eastman Kodak Co.), hydrazine sulfate (certified, Fisher Scientific Co.), sodium cyanide (certified, Fisher Scientific Co.) and chlorine gas (Matheson Co.) [15-17]. Initiator L2 was synthesized from chloroacetone (practical, Eastman Kodak Co.), sodium sulfite (reagent, J. T. Baker Chemical Co.), hydrazine sulfate, sodium cyanide and chlorine gas [15-17]. L3 was synthesized from 2-methyl 5-vinylpyridine (reagent, Aldrich Chemical Co.) and dimethyl sulfate (reagent, Aldrich Chemical Co.) [18,19]. L4 was synthesized from 2-methyl 5-vinylpyridine and ethyl bromide (reagent, Aldrich Chemical Co.) [20,21]. Sodium vinylbenzyl sulfonate (L5) was synthesized from vinylbenzyl chloride (technical, Dow Chemical

Co.) and sodium sulfite [22]. Anionic surfactant Siponate DS-10 (sodium dodecylbenzene sulfonate from Alcolac Chemical Co.) and non-ionic surfactant Triton X-100 (polyoxyethylene isooctylphenyl ether from Rohm and Haas Co.) were used as received. All the other chemicals are commonly used reagent grade chemicals. Monodisperse polybutadiene latices for the bisulfite addition studies were kindly supplied by Borg-Warner Corp. Mixed-bed ion-exchange resin (Rohm and Haas MB-3) was rinsed before use with double-distilled water until a conductivity water of specific conductance 0.3-0.5 μmho under a nitrogen atmosphere was obtained. Distilled water was used in all reactions.

III. RESULTS AND DISCUSSIONS

A. Emulsion Polymerization With L1 and L2

Weak acid groups have been found on latex particle surfaces by many investigators [5,6,10,23-25]. These weak acid groups will affect the latex surface charges in a complex way, because their dissociation is pH-dependent. It is believed that these weak acid groups are formed from a side reaction of the commonly used initiator potassium persulfate [5,27]. To avoid this complication, we synthesized the azo-type initiators L1 and L2 for use in place of persulfate. Methods we applied to synthesize these initiators are proposed by J. A. Phelisse and C. A. Quiby [15] as well as by C. G Overberger and coworkers [16,17], as follows on page 49.
Proton nmr spectrometry and elemental analysis were used to identify the individual intermediates and final products. Decomposition of one mole of these initiators forms two moles of radicals and gives off one mole nitrogen gas, as in the decomposition of AIBN. Typical recipes for polymerization using these initiators are shown in Table I.

(1) L1:

$$CH_3\text{-}\overset{O}{\overset{\|}{C}}\text{-}CH_2\text{-}\overset{O}{\overset{\|}{C}}\text{-}OEt + HO\text{-}CH_2\text{-}SO_3Na \xrightarrow[H_2O]{OH^-} CH_3\text{-}\overset{O}{\overset{\|}{C}}\text{-}\underset{\underset{SO_3Na}{|}}{\underset{CH_2}{\underset{|}{CH}}}\text{-}\overset{O}{\overset{\|}{C}}\text{-}OEt$$

$$\downarrow H^+ \quad \Delta\text{-}CO_2$$

$$CH_3\text{-}\overset{O}{\overset{\|}{C}}\text{-}CH_2\text{-}CH_2\text{-}SO_3Na$$

$$\downarrow NH_2NH_2\cdot H_2SO_4, NaCN$$

$$NaO_3S\text{-}CH_2\text{-}CH_2\text{-}\underset{CN}{\underset{|}{\overset{CH_3}{\overset{|}{C}}}}\text{-}NH\text{-}NH\text{-}\underset{CN}{\underset{|}{\overset{CH_3}{\overset{|}{C}}}}\text{-}CH_2\text{-}CH_2\text{-}SO_3Na$$

$$\downarrow Cl_2 \quad [Oxdn.]$$

$$NaO_3S\text{-}CH_2\text{-}CH_2\text{-}\underset{CN}{\underset{|}{\overset{CH_3}{\overset{|}{C}}}}\text{-}N{=}N\text{-}\underset{CN}{\underset{|}{\overset{CH_3}{\overset{|}{C}}}}\text{-}CH_2\text{-}CH_2\text{-}SO_3Na$$

(2) L2:

$$CH_3\text{-}\overset{O}{\overset{\|}{C}}\text{-}CH_2Cl + Na_2SO_3$$

$$\downarrow$$

$$CH_3\text{-}\overset{O}{\overset{\|}{C}}\text{-}CH_2\text{-}SO_3Na$$

$$\downarrow NH_2NH_2\cdot H_2SO_4, NaCN$$

$$NaO_3S\text{-}CH_2\text{-}\underset{CN}{\underset{|}{\overset{CH_3}{\overset{|}{C}}}}\text{-}NH\text{-}NH\text{-}\underset{CN}{\underset{|}{\overset{CH_3}{\overset{|}{C}}}}\text{-}CH_2\text{-}SO_3Na$$

$$\downarrow Cl_2 \quad [Oxdn.]$$

$$NaO_3S\text{-}CH_2\text{-}\underset{CN}{\underset{|}{\overset{CH_3}{\overset{|}{C}}}}\text{-}N{=}N\text{-}\underset{CN}{\underset{|}{\overset{CH_3}{\overset{|}{C}}}}\text{-}CH_2\text{-}SO_3Na$$

Table I
Typical Recipes for Preparing
Monodisperse Polystyrene Latices

Latex Code No.	LL-29A	LL-31
H_2O	1690 ml	1715 ml
Styrene	200 ml	200 ml
3.0% (w/w) L1[a]	10 ml	——
3.0% (w/w) L2[b]	——	10 ml
Temperature	65°C	65°C
Diameter	0.45μ	0.26μ

[a]L1: Azo-bis(Sodium 2-Methylbutyronitrile sulfonate)

[b]L2: Azo-bis(Sodium Isobutyronitrile sulfonate)

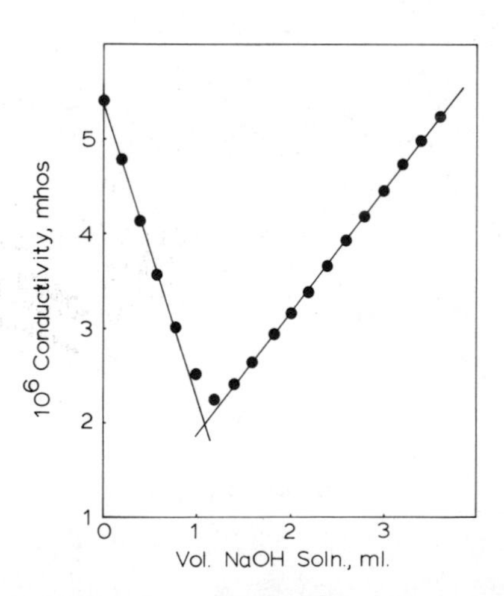

FIG. 2 Typical conductometric acid-base titration curve for latex with only strong acid surface charge.

Using these initiators, we obtained monodisperse latices with only strong acid surface charges, as indicated by the shape of the titration curves. One such curve is shown in Figure 2. With weak acid on the particle surface, one will obtain titration curves with two or more break points. Charge density in these latices is on the order of 10^{13} charges per square centimeter.

B. Emulsion Polymerization With AIBA•2HCl

Monodisperse polystyrene latices with positive surface charges were synthesized using AIBA•2HCl as initiator. Two typical recipes for polymerization using this initiator in both kettle and bottle

Table II
Typical Recipes for Preparing
Monodisperse Polystyrene Latices

Latex Code No.	LL-26C	DG-8C
H_2O	1740 ml	264 ml
Styrene	100 ml	30 ml
3.0% (w/w) AIBA•2HCl	10 ml	4 ml
0.1 N NaOH	5.6 ml	———
0.1 M KH_2PO_4	50 ml	———
Temperature	65°C	65°C
Diameter	1.53μ	0.55μ

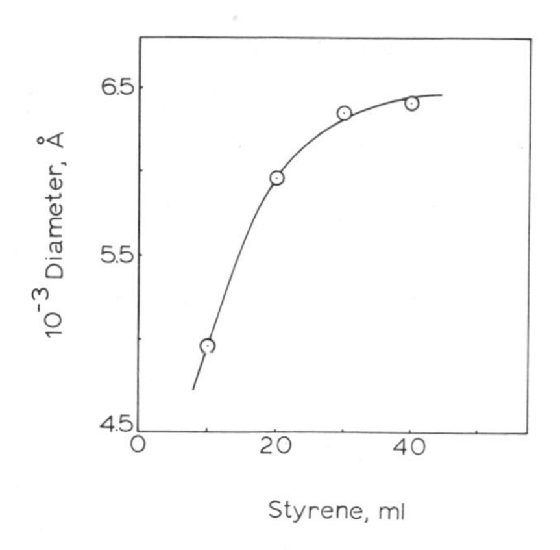

FIG. 3 Variation of particle diameter with monomer content. [AIBA•2HCl] = 5.46 x 10^{-4}M; ionic strength = 1.64 x 10^{-3}M.

reactor are shown in Table II. This initiator is less efficient than potassium persulfate, but the effects of individual ingredients on the final latex particle size in this system follow the same trend as with potassium persulfate initiator [28]. These effects are shown in Figures 3 to 5. Figure 3 shows that particle size increases with increasing monomer content when initiator concentration is constant; Figure 4 shows that particle size decreases with in-

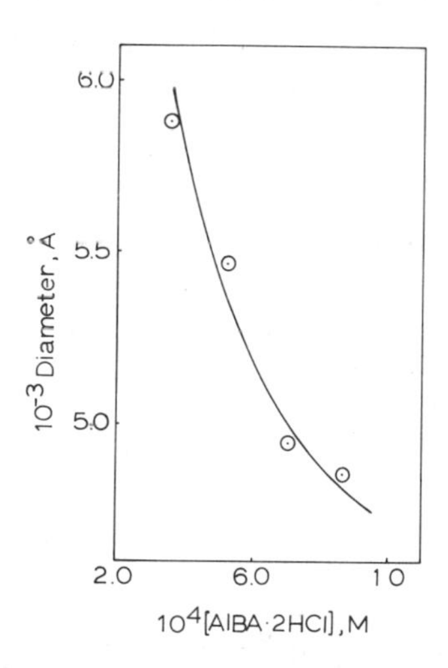

FIG. 4 Variation of particle diameter with initiator concentration. [styrene] = 0.870M; ionic strength = 2.46 x 10^{-3}M.

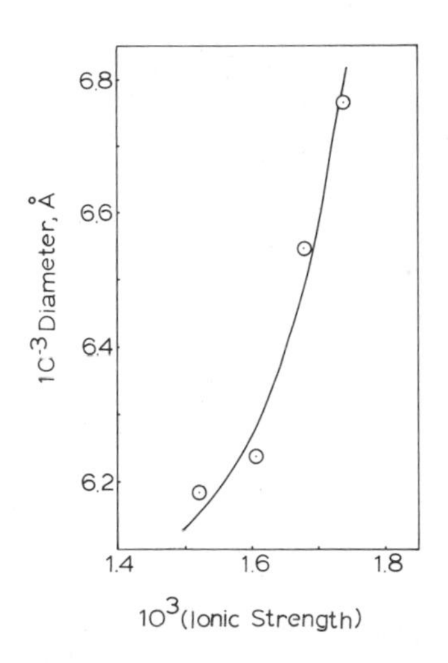

FIG. 5 Variation of particle diameter with ionic strength. [AIBA•2HCl] = 5.46 x 10^{-4}M; [styrene] = 0.870M.

creasing initiator concentration when both monomer content and ionic strength are kept constant; and Figure 5 shows the effect of ionic strength. Particle size increases with increasing ionic strength at constant monomer and initiator concentration. Ionic strength was increased by adding a neutral electrolyte such as sodium chloride to the system.

Surface charges in this system come from the weakly basic amidinium ions of the initiator; therefore, charge on the particle will be pH-dependent. At high pH, amidinium ion changes to amidine and the charge disappears; as a result, the latex becomes unstable. We determined surface charge densities of these latices by adding nonionic surfactant to the latex before conductometric titration to increase its stability. The surface charge density is on the order of 10^{13} charges/cm^2, which is a typical surface charge density for an emulsifier-free latex.

C. Emulsion Polymerization With Ionic Comonomers L3, L4 and L5

Besides the advantages of increased solids content of the latex and wider range of particle sizes that can be synthesized, ionic

Table III
Typical Recipe for Preparing
Monodisperse Polystyrene Latex

Latex Code No.	LL-84E
H_2O	90 ml
Styrene	90 ml
0.33% (w/w) L5[a]	90 ml
1.0% (w/w) $K_2S_2O_8$	30 ml
Temperature	65°C
Diameter	0.28μ

[a]L5 = Sodium Vinylbenzyl Sulfonate

comonomer can independently control surface charge density [10]. In an emulsifier-free system, surface charge can be varied by varying the ionic comonomer concentration while keeping the initiator concentration constant. One of the anionic comonomers we used recently is L5; a typical recipe using this ionic comonomer is shown in Table III. Variation of surface charge density using this ionic comonomer is shown in Table IV. It is clear from the table that a less than 4-fold increase in ionic comonomer concentration gives a more than 5-fold increase in strong acid surface charge density. The effects of other variables on this system are shown in Figures 6 to 9. We also synthesized two cationic comonomers, L3 and L4, to prepare latices with positive surface charges. In Table V, two typical recipes using these ionic comonomers are shown. Systems with these ionic comonomers behave similarly to the anionic comono-

Table IV

Effect of Anionic Comonomer Concentration on Particle Size and Surface Charge Density[a]

[NaVBS] mol/ℓ	Particle Diameter μ	Strong Acid Charge/Å^2	Weak Acid Charge/Å^2
7.56×10^{-4}	0.84	2.95×10^{-4}	6.61×10^{-3}
1.89×10^{-3}	0.58	1.17×10^{-3}	5.32×10^{-3}
3.03×10^{-3}	0.46	1.57×10^{-3}	6.77×10^{-3}

[a] $[K_2S_2O_8] = 3.70 \times 10^{-3}$ mol/ℓ
[styrene] = 2.61 mol/ℓ
Ionic strength = 2.36×10^{-2}

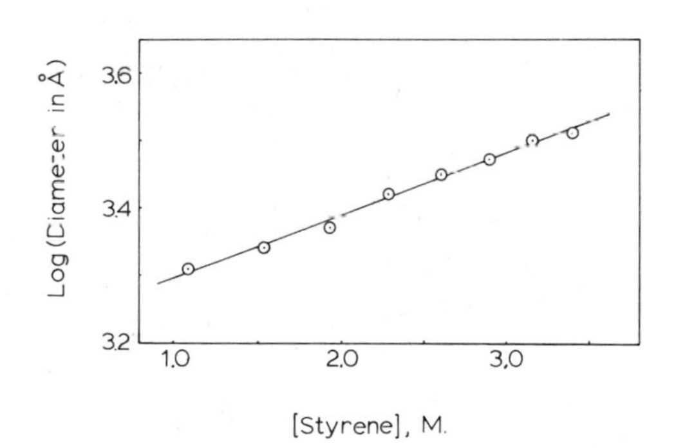

FIG. 6 Variation of particle diameter with monomer content. $[K_2S_2O_8] = 5.28 \times 10^{-3}$M; [NaVBS] = 6.49×10^{-3}M; ionic strength = 2.23×10^{-2}M.

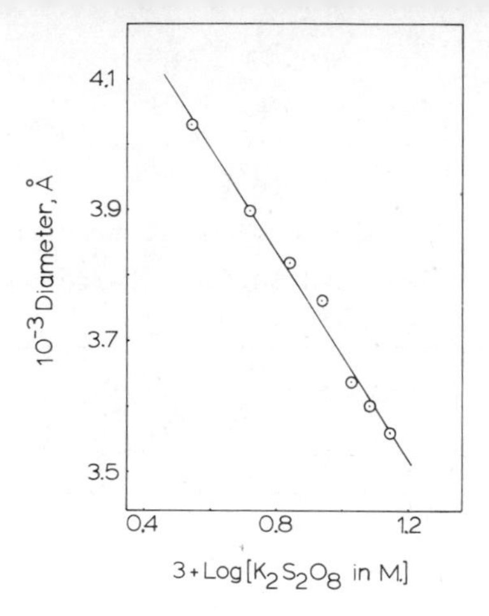

FIG. 7 Variation of particle diameter with initiator concentration. [NaVBS] = 6.49 x 10^{-3}M; [styrene] = 2.61M; ionic strength = 4.88 x 10^{-2}M.

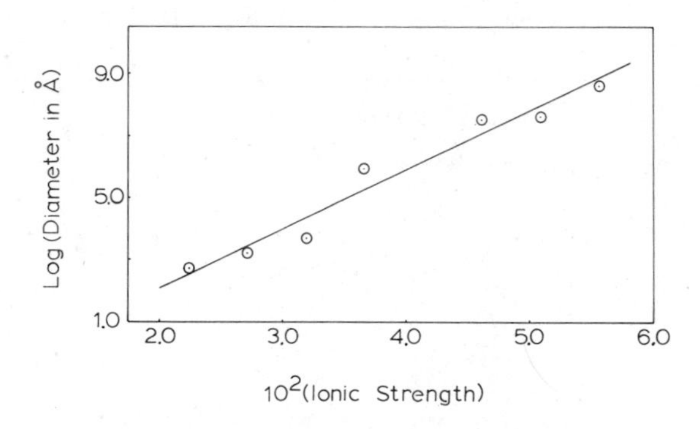

FIG. 8 Variation of particle diameter with ionic strength. [$K_2S_2O_8$] = 5.28 x 10^{-3}M; [NaVBS] = 6.49 x 10^{-3}M; [styrene] = 2.61M.

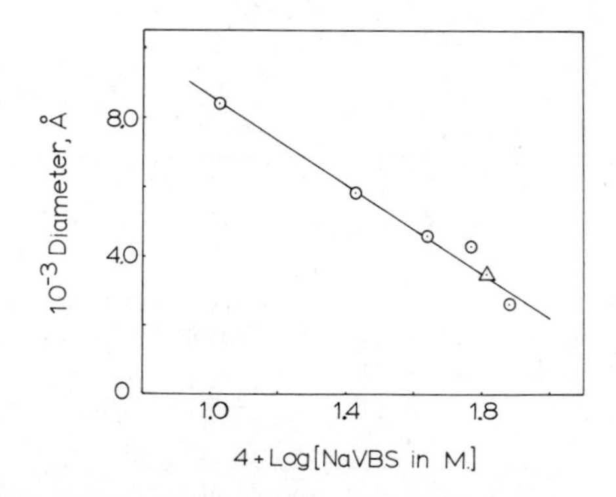

FIG. 9 Variation of particle diameter with ionic comonomer concentration. [$K_2S_2O_8$] = 5.28 x 10^{-3}M; [styrene] = 2.61M; ionic strength = 2.83 x 10^{-2}M. Different points refer to different batches.

Table V
Typical Recipes for Preparing
Monodisperse Polystyrene Latices

Latex Code No.	LL-36E	LL-45D
H_2O	150 ml	150 ml
Styrene	90 ml	75 ml
1% (w/w) AIBA•2HCl[a]	30 ml	30 ml
1% (w/w) L3[b]	____	30 ml
1% (w/w) L4[c]	30 ml	____
Temperature	65°C	65°C
Diameter	0.24μ	0.26μ

[a]AIBA•2HCl: Azo-bis(isobutyramidine hydrochloric acid)
[b]L3: 1,2-Dimethyl 5-vinylpyridinium methyl sulfate
[c]L4: 1-Ethyl 2-methyl 5-vinylpyridinium bromide

mer systems we had reported earlier [10], as well as those we have worked with lately, as shown in Figures 10 to 13. Figure 10 shows that particle size increases with increasing monomer content; Figure 11 shows that particle size decreases with increasing initiator concentration; Figure 12 shows particle size increases with increasing ionic strength; Figure 13 shows the effect of ionic comonomer concentration. Particle size decreases first, and then increases with increasing ionic comonomer concentration.

In systems with different concentrations of cationic comonomer, variation of surface charge density was also observed as shown in

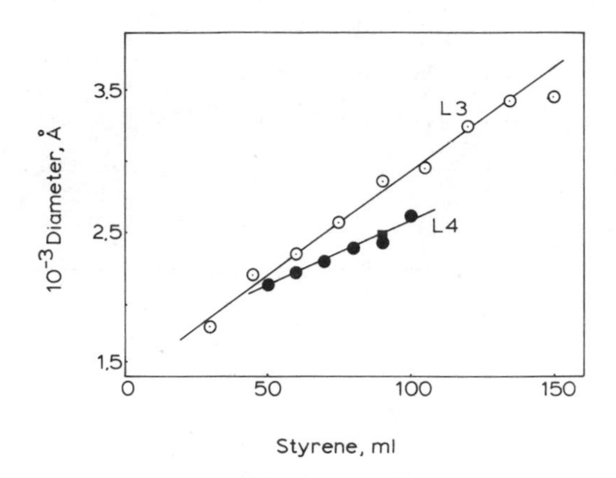

FIG. 10 Variation of particle diameter with monomer content in systems using L3 (open symbols) and L4 (filled symbols) as ionic comonomers. L3:[AIBA•2HCl] = 5.27 x 10^{-3}M; [L3] = 5.83 x 10^{-3}M; ionic strength = 2.16 x 10^{-2}M. L4:[AIBA 2HCl] = 5.27 x 10^{-3}M; [L4] = 6.26 x 10^{-3}M; ionic strength = 2.21 x 10^{-2}M.

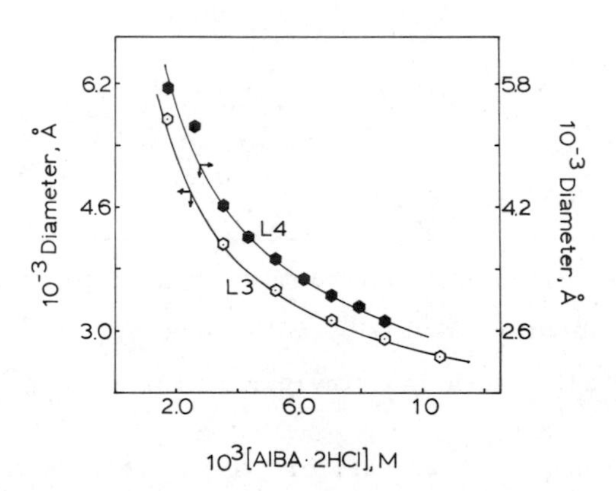

FIG. 11 Variation of particle diameter with initiator concentration in systems with L3 (open symbols) and L4 (filled symbols) as ionic comonomers. L3:[L3] = 5.83 x 10^{-3}M; [styrene] = 2.29M; ionic strength = 3.75 x 10^{-2}M. L4:[L4] = 6.26 x 10^{-3}M; [styrene] = 2.61M; ionic strength = 3.26 x 10^{-2} M.

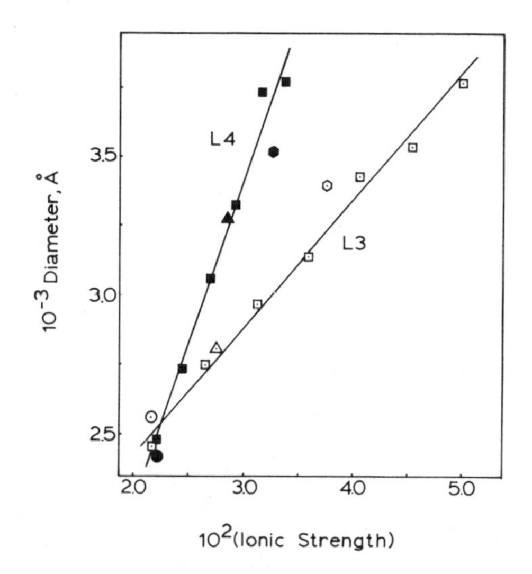

FIG. 12 Variation of particle diameter with ionic strength in system with L3 (open symbols) and L4 (filled symbols) as ionic comonomers. L3 = [AIBA·2HCl] = 5.27 x 10^{-3}M; [L3] = 5.83 x 10^{-3}M; [styrene] = 2.29M. L4: [AIBA·2HCl] = 5.27 x 10^{-3}M; [L4] = 6.26 x 10^{-3}M; [styrene] = 2.61M.

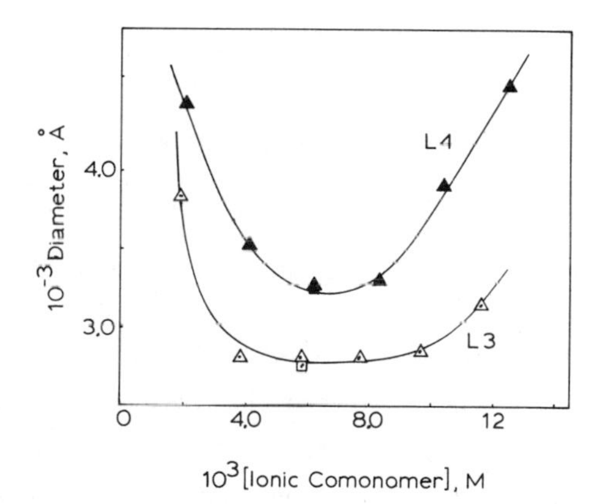

FIG. 13 Variation of particle diameter with ionic comonomer concentration in systems with L3 (open symbols) and L4 (filled symbols) as ionic comonomers. L3: [AIBA·2HCl] = 5.27 x 10^{-3}M; [styrene] = 2.29M; ionic strength = 2.75 x 10^{-2}M. L4: [AIBA·2HCl] = 5.27 x 10^{-3}M; [styrene] = 2.61M; ionic strength = 2.83 x 10^{-2}M.

Table VI

Effect of Cationic Comonomer Concentration on Particle Size and Surface Charge Density[a]

[Ionic Comonomer] mol/ℓ		Particle Diameter u	Strong Base Charge/Å^2	Weak Base Charge/Å^2
L3[b]:	1.43×10^{-3}	0.38	9.98×10^{-4}	2.97×10^{-3}
	4.29×10^{-3}	0.28	1.10×10^{-3}	3.60×10^{-3}
	7.15×10^{-3}	0.28	2.09×10^{-3}	3.73×10^{-3}
L4[c]:	1.46×10^{-3}	0.44	4.60×10^{-4}	2.75×10^{-3}
	4.38×10^{-3}	0.33	1.39×10^{-3}	2.40×10^{-3}
	7.31×10^{-3}	0.39	2.05×10^{-3}	4.25×10^{-3}

[a] [AIBA•2HCl] = 5.27×10^{-3} mol/ℓ

[Styrene] = 2.30 mol/ℓ(L3), 2.61 mol/ℓ (L4)

Ionic strength = 2.75×10^{-2} (L3), 2.83×10^{-2} (L4)

[b] L3: 1,2-Dimethyl 5-Vinylpyridinium Methyl Sulfate

[c] L4: 1-Ethyl 2-Methyl 5-Vinylpyridinium Bromide

Table VI. A 5-fold increase in ionic comonomer L3 results in a 2-fold increase in strong base surface charges, whereas in the system with L4 as ionic comonomer, a 5-fold increase in concentration results in a 3-fold increase in strong base surface charges.

D. Surface Charge Modification of Polybutadiene Latices

One of the surface charge modification reactions we are interested in is formulated as follows:

$$HSO_3^- \rightarrow \cdot SO_3^- \quad (1)$$

$$\cdot SO_3^- + \;>C{=}C< \;\rightarrow \cdot\overset{|}{\underset{|}{C}}-\overset{|}{\underset{|}{C}}SO_3^- \quad (2)$$

$$\cdot\overset{|}{\underset{|}{C}}-\overset{|}{\underset{|}{C}}SO_3^- + HSO_3^- \rightarrow H\overset{|}{\underset{|}{C}}-\overset{|}{\underset{|}{C}}SO_3^- + \cdot SO_3^- \quad (3)$$

When a double bond is treated with bisulfite ion in aqueous phase in the presence of air, oxygen or peroxide, addition of the bisulfite ion to the double bond occurs, which results in introducing a charged group [29,30]. Some preliminary work has been done with monodisperse polybutadiene latices. The result is shown in Figure 14. It is clear that increase in surface charge arises mainly from weak acid; contribution of the strong acid group is less than 10%. A possible reason for this result is that for an ion-radical to approach and enter the polymer surface, an appreciable potential energy barrier caused by electrostatic repulsion must be overcome. Therefore, the sulfite ion-radical would have little tendency to

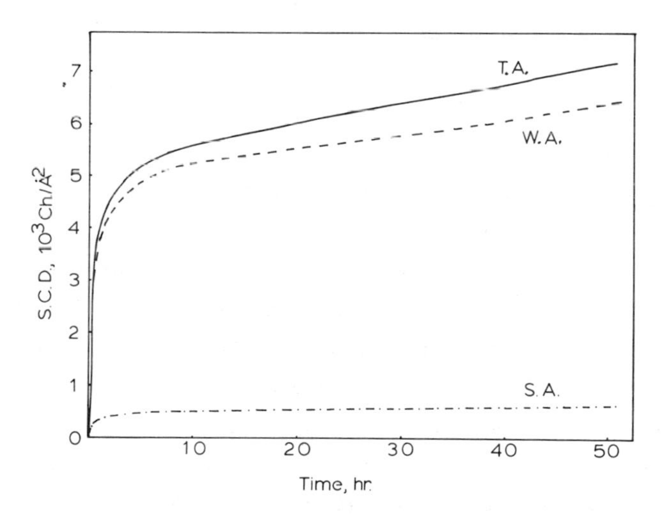

FIG. 14 Surface charge density (S.C.D.) vs. reaction time curves for addition of sodium bisulfite to polybutadiene latex system. S.A.: strong acid; W.A.: weak acid; T.A.: total acid. $[NaHSO_3] = 5.63 \times 10^{-2}M$.

migrate to the particle/water interface. We suspect that sulfite ion-radicals react with water molecules to form hydroxyl radicals. The potential energy barrier would not prevent the entry of hydroxyl radicals, because they are neutral species. As a result, an appreciable number of hydroxyl groups will combine with the polymer particles. Trace amounts of heavy metal ion in the system may catalyze the oxidation of hydroxyl groups and result in the formation of carboxyl groups on the particle surface.

E. Surface Charge Modification of Poly(vinylbenzyl chloride) Latices

When benzyl chloride is treated with sodium sulfite at 65°C, replacement of the chlorine atom by the sulfite ion occurs, which results in introducing a charge group [22]. The reaction is illustrated as follows:

$$-CH\text{-}CH- \;(C_6H_4)\; CH_2Cl \xrightarrow[65^\circ C]{Na_2SO_3} -CH\text{-}CH- \;(C_6H_4)\; CH_2SO_3^-Na^+ \quad + \quad NaCl$$

We have initiated work with monodisperse poly(vinylbenzyl chloride) latex, and obtained very encouraging results.

Because benzyl chloride hydrolyzes readily at high temperature [31-33], these latices were synthesized at room temperature using persulfate/bisulfite/Fe(II) redox initiator and mixed anionic and nonionic emulsifier. Typical recipes are shown in Table VII.

The results of conductometric acid-base titration of these latices indicate that only strong acid groups are present on the latex particle surface. This method thus introduces a simple way to synthesize latices with only strong acid surface charges; low polymerization temperature in this system may be the reason for this result.

Nonionic surfactant was added to the latex to help in stabilizing the latex particles, after which sodium sulfite solution was

Table VII

Typical Recipes for Preparing

Monodisperse Poly(vinylbenzyl chloride) Latices

Latex Code No.	LL-101	LL-107
H_2O	150 ml	45 ml
10.0% Triton X-100	40 ml	16.4 ml
1.0% Siponate DS-10	40 ml	16.4 ml
0.1N KOH	60 ml	24.6 ml
0.01% $FeSO_4 \cdot 7H_2O$	10 ml	4.1 ml
Vinylbenzyl Chloride	240 ml	100 ml
CH_3NO_2	0.27 g	0.11 g
Divinylbenzene	10 ml	——
1.0% $NaHSO_3$	40 ml	16.4 ml
3.0% $K_2S_2O_8$	20 ml	——
1.0% $K_2S_2O_8$	——	24.6 ml
Temperature	25°C	30°C
Diameter	0.19μ	0.18μ

added at room temperature. The temperature of the medium was then raised to 65°C in a constant temperature water bath. Latex samples were withdrawn at different times to measure the extent of the reaction. Substitution vs. time is shown in Figure 15 and Table VIII. An increase in surface charge density of almost 20-fold occurred in less than 50 hours. Electron microscopy showed no detectable change in particle size. Figure 15 also shows that pH dropped when the reaction time increased, as a consequence of the substitution reaction.

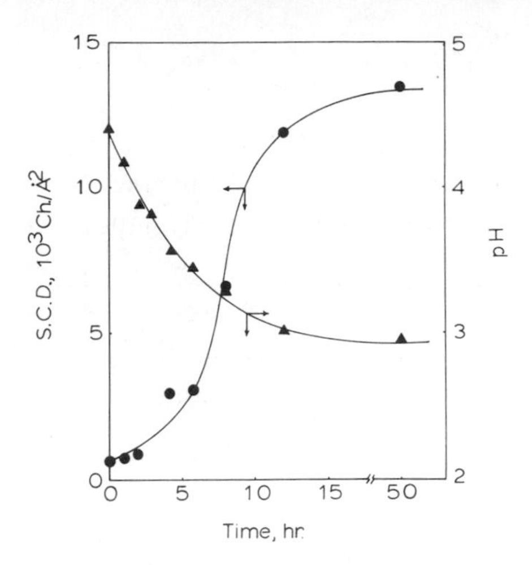

FIG. 15 Variation of surface charge density (S.C.D.) and pH with time in reaction of poly(vinylbenzyl chloride) latex with sodium sulfite at 65°C. $[Na_2SO_3]$ = 2.88 x 10^{-2}M.

Table VIII

Reaction of Poly(vinylbenzyl chloride) Latices With Sodium Sulfite[a]

Sample #	Rx'n Time hr.	Surface Charge Density 10^4 Charge/$Å^2$	pH
1	0	6.85	4.40
2	1.0	7.74	4.17
3	2.0	8.00	3.88
4	3.0	14.5	3.82
5	4.25	29.2	3.56
6	5.75	30.8	3.45
7	8.0	65.9	3.28
8	12.0	118	3.00
9	47.25	135	2.95

[a] $[Na_2SO_3]$ = 2.88 x 10^{-2} M

Temperature = 65°C

IV. CONCLUSIONS

Two of the most attractive routes to control of latex particle surface charge produced undesirable weak acid surface groups along with the strong. These are (1) incorporation of ionic comonomer in a persulfate-initiated emulsifier-free recipe, and (2) addition of bisulfite across the surface unsaturation in a polybutadiene latex. Success in producing only strongly acidic groups was achieved by two different routes. The first was replacement of persulfate, in emulsifier-free recipes containing ionic comonomer, by a sulfonated azo-type initiator. The second successful route started with poly-(vinylbenzyl chloride) latex, prepared at room temperature with a redox initiator, and added sodium sulfite to replace chlorine on the surface by sulfonate groups. This latter method can produce a 20-fold variation of surface charge, without altering particle diameter.

Monodisperse latices with positive surface charges were produced using cationic azo-type initiators and/or cationic comonomers.

V. ACKNOWLEDGMENTS

This work was partially supported by a grant from the Public Health Service. One of the authors (L.-J.L.) held a Paint Research Institute Fellowship during the course of this investigation.

REFERENCES:

1. K. J. Mysels, Introduction to Colloid Chemistry, 4th Printing, Interscience Inc., N. Y. (1967).

2. J. Stone-Masui and A. Watillon, J. Colloid and Interface Sci., 28, 187 (1968).

3. I. M. Krieger and M. Eguiluz, Trans. Soc. Rheol., 20:1, 29 (1976).
4. W. V. Smith, J. Am. Chem. Soc., 70, 3695 (1948).
5. H. J. van den Hul and J. W. Vanderhoff, Br. Polym. J., 2, 121 (1970).
6. J. Stone-Masui and A. Watillon, J. Colloid and Interface Sci., 52, 479 (1975).
7. W. C. Wu, Ph.D. Thesis, Lehigh Univ., (1977).
8. R. M. Fitch and W. T. McCarvill, Preprint of 173rd ACS Colloid and Surface Chemistry National Meeting, March (1977).
9. J. R. McCracken and A. Datyner, J. Appl. Polym. Sci., 18, 3365 (1974).
10. M. S. Juang and I. M. Krieger, J. Polym. Sci., Polym. Chem. Ed., 14, 2089 (1976).
11. J. Hen, J. Colloid Interface Sci., 49, 425 (1974).
12. H. Ono, E. Jidai and K. Shibayama, Colloid Polym. Sci., 255, 105 (1977).
13. F. S. Chan and D. A. I. Goring, Canadian J. Chem., 44, 725 (1966).
14. P. Swaraj and R. Bengt, Macromolecules, 9, 337 (1976).
15. J. A. Pheliss and C. A. Quiby, U.S.P. 3,161,630 (1964).
16. C. G. Overberger, M. T. O'Shaughnessy and H. Shalit, J. Am. Chem. Soc., 71, 2661 (1949).
17. C. G. Overberger, P. Huang and M. Berenbaum, Organic Syntheses Collective Volume 4, 274 (1963).
18. W. P. Shyluk, J. Polym. Sci., Part A, 2, 2191 (1964).

19. J. C. Salamone, D. F. Bardoliwalla, E. J. Ellis, S. C. Israel and A. W. Wisniewski, Applied Polym. Symposium, No. 26, 309 (1975).

20. J. Gregory and I. Sheiham, Br. Polym. J., 6, 47 (1974).

21. D. R. Kasper, Ph.D. Thesis, California Institute of Technology (1971).

22. G. D. Jones, U.S.P. 2,909,508 (1959).

23. J. N. Shaw, J. Polym. Sci., Part C, No. 27, 237 (1969).

24. I. M. Kolthoff and I. K. Miller, J. Am. Chem. Soc., 73, 3055 (1951)

25. R. H. Ottewill and J. N. Shaw, Kolloid Z. Z. Polym., 218, 34 (1967).

26. J. D. Smitham, D. V. Gibson and D. H. Napper, J. Colloid Interface Sci., 45, 211 (1973).

27. F. A. Bovey, I. M. Kolthoff, A. I. Medalia and E. J. Meehan, Emulsion Polymerization, Interscience, New York (1955).

28. J. W. Goodwin, J. Hearn, C. C. Ho and R. H. Ottewill, Colloid Polym. Sci., 252, 404 (1974).

29. M. S. Kharasch, E. M. May and F. R. Mayo, Chem. & Ind., 57, 774 (1938).

30. M. S. Kharasch, E. M. May and F. R. Mayo, J. Org. Chem., 3, 175 (1938).

31. W. G. Llord and J. F. Vitkuske, J. Appl. Poly. Sci., 6, No. 24, S58 (1962).

32. W. G. Llord and T. E. Durocher, J. Appl. Poly. Sci., 7, 2025 (1963).

33. J. G. Abramo and E. C. Chapin, USP 3,069,399 (1962).

DISCUSSION

A. H. Herz (Eastman Kodak Co)

For the core where the co-monomer is ionic, one expects that in addition to the ionizally modified latex the solution phase may contain ionic oligomer. Since Ottewill, et al., showed that a purified latex with few ionic surface charges exhibited surface tensions close to that of water, it would be instructive to know the surface tension of the latex described in this work. If the surface tension is low, this could indicate the presence of such ionic solution polymers which would work.

Author We have cleaned our latices with ion exchange resin. The result does show surface tension close to that of water.

A. H. Herz (Eastman Kodak Co)

Early work by Ottewill, et al., demonstrated that polystyrene latices containing covalently bound anionic groups could be purified to give surface tensions close to those of water. Purification is frequently difficult; work in our Research Laboratories with low levels of anionic co-monomers often yielded latices with surface tensions in the 50-60 dyne range despite extensive dialysis. Presumably, this surface activity was due to dissolved anionic oligomers; their presence during acid/base titrations would be expected to interfere with latex surface charge determinations. Accordingly it would be instructive to learn how the authors established purity of the latex continuous phase. For example, did these latices exhibit the surface tension of water?

Author By using mix-bed ion-exchange resin (Rohm and Haas MD-3) which was cleaned before use, we got surface tension close to that of water.

E. D. Goddard (Union Carbide Corp)

Following Dr. Herz's question, I wonder what surface tension lowering one would expect from an ionic (homo) oligomer. I would suspect it could be very small, depending of course on the particular ionic monomer.

Author If the molecular weight of the oligomer is high, it should not lower the surface tension appreciably.

J. W. Vanderhoff (Emulsion Polymers Institute, Lehigh University)

I would like to comment on the removal of water-soluble polymers from the aqueous serum by ion-exchange. In a study of styrene-butadiene-acrylic acid copolymer (Polymer Preprints (1), 1975), we found that a low-molecular-weight carboxyl-containing polymer was removed quantitatively from the aqueous phase by ion-exchange. Moreover, the surface tension of these latexes decreased with increasing acrylic acid concentration, from 72 dynes/cm without acrylic acid to about 50 dynes/cm for 5% acrylic acid, indicating that these polymers are surface-active.

John Hen (Uniroyal Chemical)

Verification of comments of author on his ability to obtain high surface tension latexes in range of that of water from latexes prepared with ionic comonomers. My own experience as well as that of other workers is that ion-exchange is a highly effective means of removing unbound oligomer polyelectrolyte, which should result in a latex of surface tension like that of water if the ion-exchange was carried out properly.

W. W. White (Uniroyal Chemical)

Oligomeric surfactants (Polywet) present in latexes prepared where they are the sole emulsifier give latexes with surface tensions of 70 dynes/cm. This shows that low surface tension is not a good indication of the presence of oligomeric materials in the aqueous phases.

Author The surface tension of our latices after ion-exchange is close to that of water, i.e., ca. 71 dynes/cm.

R. O. James (Stanford University)

What techniques did you use to distinguish the strong and weak acid ionisable sites of your latexes? Did you use potentiometric titration?

Author We use conductometric titration to distinguish the strong and weak acid groups. Addition of a small amount of salt like sodium chloride does help to get a better measurement of the end point.

SURFACE CHARGE DENSITY AND ELECTROPHORETIC MOBILITY OF MONODISPERSE POLYSTYRENE LATEXES

W. C. Wu, M. S. El-Aasser, F. J. Micale,
and J. W. Vanderhoff

Emulsion Polymers Institute
Lehigh University
Bethlehem, Pennsylvania

ABSTRACT

A new experimental approach has been developed to characterize the surface of monodisperse polystyrene latexes. By a combination of conductometric and potentiometric titration, conductometric, and electrophoretic techniques, the apparent degree of dissociation of the surface sulfate groups of a cleaned 357nm-diameter monodisperse polystyrene latex in different ionic forms (H^+, Na^+, Ba^{++}) was found to range from 0.063 for the H^+ form to 0.7 for the Na^+ form. The conductometric and potentiometric results as a function of electrolyte concentration and polymer concentration can be interpreted qualitatively in terms of the variation of the apparent degree of dissociation of the sulfate groups and the interfacial ion-exchange. When an electrolyte with a different cation is added to the cleaned latex, the sudden increase in the concentration of cations and their associated anions in the aqueous bulk phase of the latex upsets the equilibrium state existing between the ions in the surface and the aqueous bulk phase. Interfacial ion exchange between the two phases takes place until a new equilibrium state is reached. Thus, when an ion-exchanged polystyrene latex is titrated with sodium hydroxide, two types of interfacial reactions---neutralization and ion exchange---take place concurrently.

INTRODUCTION

Monodisperse polystyrene latexes have been widely used as model colloids for colloidal studies because their particles are rigid, spherical, and uniform in size, and they can be easily prepared. Recently, a method was developed to clean and characterize these latexes (1-6), which comprises ion exchange with mixed resins to remove the adsorbed emulsifier and solute electrolyte and to convert the surface groups to the H^+ form, followed by conductometric titration with base. Thus, the type and number of the surface groups of the latex can be determined. Studies of the characterization of these latexes (3, 7, 8) have shown that they are not ideal model colloids as originally thought; the nature of the surface charge, which is a function of the polymer endgroups on the latex particle surface, depends upon the polymerization recipes and the conditions of preparation. The colloidal properties of the polystyrene latexes prepared in different ways and under different conditions may vary considerably. Recently, it has been found that the effective surface charge density of cleaned polystyrene latex particles does not correspond to the surface density of the sulfate endgroups, and the latexes behave as if only part of the sulfate groups were dissociated (2, 4, 9, 10). The nature of this phenomenon must be understood to interpret the colloidal behavior of these latexes when they are used as model colloids. This paper describes the development of experimental methods to investigate this phenomenon and the characterization of the latex surface properties as a function of the electrolyte concentration and type of counterion.

EXPERIMENTAL DETAILS

All chemicals used were of A.R. grade. The distilled water used was double-distilled and de-ionized.

The monodisperse polystyrene latexes to be studied were first characterized using the ion-exchange-and-conductometric titration techniques described earlier (3, 10) to determine quantitatively the polymer endgroups bound to the particle surface. Only those latexes which were found to have no surface carboxyl groups were used for further investigation.

The pH and conductivity of the cleaned monodisperse latexes were measured at different polymer concentrations. A Fisher Accumet Model 230 pH meter was used to measure the pH, and a conductometric apparatus capable of continuous recording was used to measure the conductivity (11).

The serum phase of the latex was separated by first freeze-coagulating the cleaned latex in a polyethylene bottle and then filtering through quartz wool. The quality of the serum was examined by conductometric titration with 0.02N sodium hydroxide.

Ion-exchanged latexes in the H^+ form were converted to the Na^+, Ba^{++}, or other cation forms by adding, respectively, the exact amount of sodium hydroxide, barium hydroxide, or other hydroxide solution required to neutralize the H^+ ions present in the latex (determined from conductometric titration results).

The electrophoretic mobilities of the latexes in the H^+, Na^+, and Ba^{++} forms were measured at different polymer concentrations and different electrolyte concentrations using a Micromeritics Model 120 Electrophoretic Mass Transport Analyzer. This instrument measures the increase in latex particle concentration in a small cell next to the electrode of opposite charge; this increase in concentration is then used to calculate the electrophoretic mobility. The cell and reservoir chamber were filled with the latex dispersion of known polymer concentration and conductivity and kept under constant rotation to prevent any settling of the particles. An electric field was applied across the electrodes, causing a migration of the particles into the cell and the solvent and counterions into the reservoir chamber. The electrophoresis was continued for a period of time required to give a significant change in particle concentration. A modified experimental technique was used to determine the electrophoretic mobilities of the latexes. Instead of measuring the weight difference of the cell before and after electrophoresis (which would be very small for particles having almost the same density as the liquid medium), the weight difference of the total solids in the cell was measured. Before and after each electrophoresis experiment, the latex in the cell was washed out quantitatively into weighing dishes and evaporated to constant weight.

The electrophoretic mobility was calculated from the measured experimental parameters: the weight of particles W_p (gm) which enter the cell (it is presumed that particles migrate into the cell rather than from it); the conductivity λ ($ohm^{-1}cm^{-1}$) of the latex; the time $\underline{t}$ (sec) of the electrophoresis run; the current $\underline{i}$ (ampere); solids concentration $\underline{c}$ (gm/cc); the solids volume fraction $\emptyset$. The mass of the particles $\underline{w}_p$ entering the cell per unit time $\underline{t}$ due to electrophoretic migration can be written as

$$w_p/t = VAc \tag{1}$$

where $\underline{V}$ is the velocity of migration, $\underline{A}$ the cross-sectional area of the entry passageway to the cell, and $\underline{c}$ the mass concentration of the particles.

Under steady flow conditions, the particle velocity is given by

$$V = E\,U_e - V_1 \tag{2}$$

where $\underline{E}$ is the potential gradient, $\underline{U}_e$ the electrophoretic mobility, and $\underline{V}_1$ the liquid velocity from the cell. Equation 2 can be rearranged to

$$U_e = (V + V_1)/E \tag{3}$$

The liquid velocity $\underline{V}_1$ is determined by the rate at which liquid is displaced divided by the cross-sectional area of the entry passageway to the cell and is given by

$$V_1 = V\emptyset/(1 - \emptyset) \tag{4}$$

The potential gradient $\underline{E}$ across the cell entranceway is given by

$$E = i/(\lambda A) \tag{5}$$

Substituting Equation 4 into Equation 3 and rearranging gives

$$U_e = [V/(1-\emptyset)]/E \tag{6}$$

Rearranging Equation 1 and substituting it into Equation 6 gives

$$U_e = [W_p/(Act(1 - \emptyset))]/E \tag{7}$$

Substituting Equation 5 into Equation 7 gives the expression

$$U_e = (W_p\lambda)[ict(1 - \emptyset)] \tag{8}$$

in which $\underline{c}$ can be measured experimentally while $\emptyset$ can be calculated if the density ρ_p of the particles is known.

From the conductivity, conductometric titration, and electrophoretic mobility measurements, the net surface charge of the latex particles subjected to an electric field and the corresponding values of the apparent degree of dissociation of the functional endgroups can be determined.

The interaction of different cations at the latex particle surface was studied by conductometric titration of the ion-exchanged latexes in the H^+ form with sodium hydroxide as a function of electrolyte concentration; the phenomenon was also studied by following the variation of conductivity of the ion-exchanged latexes in the H^+ form with different electrolytes such as potassium chloride (K^+) and barium chloride (Ba^{++}).

EXPERIMENTAL RESULTS

To study the effect of electrolyte concentration on the surface properties of polystyrene latexes, a 234nm-diameter monodisperse polystyrene latex was titrated conductometrically with 0.02N sodium hydroxide at different potassium and barium chloride concentrations.

Figure 1 shows that conductometric titration of the latex as a function of potassium chloride concentration in the range 0 - 1.5×10^{-5}M. The linear descending legs (due to sulfate groups) increase in slope and deviate slightly from linearity with increasing potassium chloride concentration. The linear ascending legs (due to excess sodium hydroxide) have about the same slopes. The endpoints are the same within experimental error.

Figure 2 shows the data of Figure 1 as the variation of initial conductivity with potassium chloride concentration. Line 2 shows the values expected assuming the conductivities are additive and Line 1 shows the experimental values. The experimental values are greater, i.e., the potassium chloride increased the conductivity of the latex by a factor greater than its own conductivity.

Figure 3 shows the conductometric titration of the same latex as a function of barium chloride concentration in the range 0–0.75 x 10^{-5}M. The increase in slope of the linear descending legs with increasing barium chloride concentration is greater than with potassium chloride.

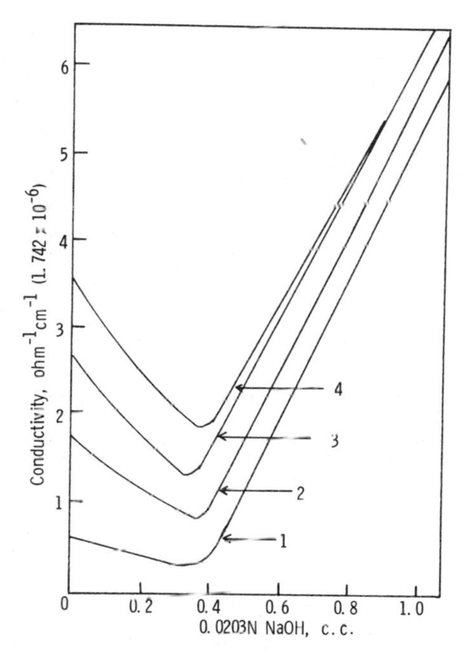

Figure 1. Conductometric titration of ion-exchanged 234nm-diameter monodisperse polystyrene latex as a function of potassium chloride concentration: 1. none; 2. 0.5 x 10^{-5}M; 3. 1.0 x 10^{-5}M; 4. 1.5 x 10^{-5}M.

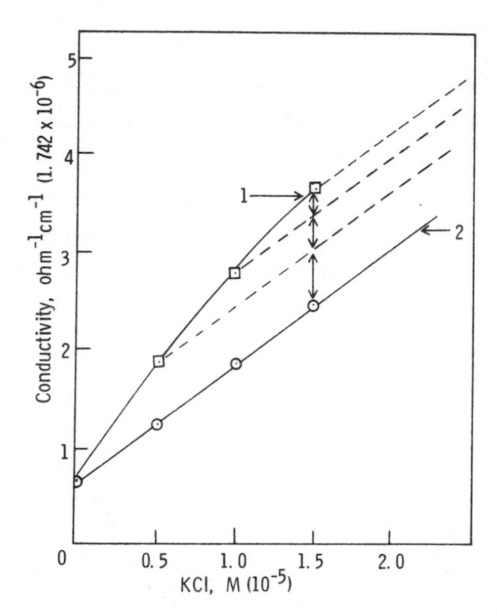

Figure 2. Variation of conductivity of ion-exchanged 234nm-diameter monodisperse polystyrene latex with potassium chloride concentration: 1. experimental values; 2. values calculated assuming additive conductivities and no interactions.

Figure 4 shows the data of Figure 3 as the variation of initial conductivity with barium chloride concentration. The extra conductance is greater for barium chloride than for potassium chloride.

At the same time, the initial pH of the same latex (before any addition of sodium hydroxide) was measured as a function of potassium chloride and barium chloride concentrations. Figure 5 shows that the H^+ ion concentrations increase with increasing potassium chloride and barium chloride concentrations. The increase in H^+ ion concentration in the bulk phase with increasing barium chloride concentration is greater than with potassium chloride.

The H^+ ion concentrations in the aqueous bulk phase (dissociated) at different potassium chloride and barium chloride concentrations can be estimated from the difference between the initial conductivity values and the minimum values (at the end points) of the conductometric titration curves. The ratio of this estimated dissociated H^+ ion concentration to the expected H^+ ion concentration (assuming complete dissociation of sulfate groups) is called the apparent

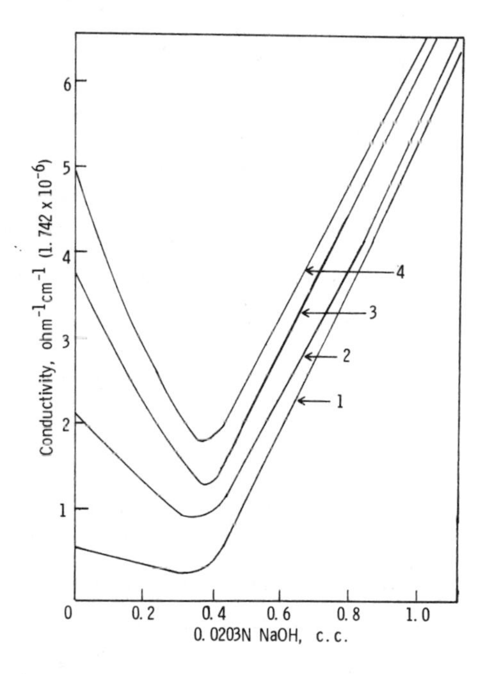

Figure 3. Conductometric titration of ion-exchanged 234nm-diameter monodisperse polystyrene latex as a function of barium chloride concentration: 1. none; 2. 0.25×10^{-5}M; 3. 0.50×10^{-5}M; 4. 0.75×10^{-5}M.

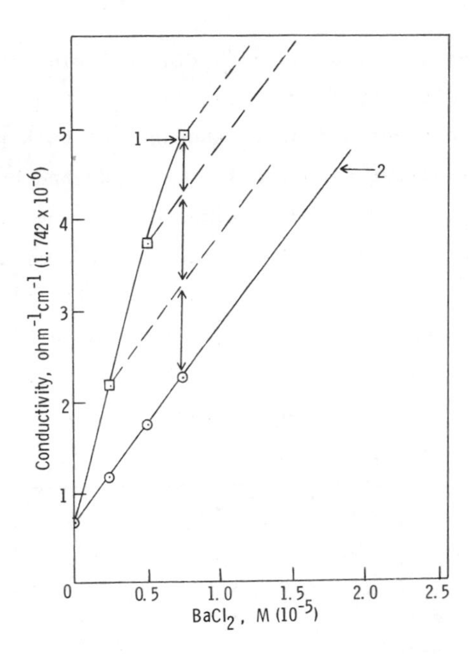

Figure 4. Variation of conductivity of ion-exchanged 234nm-diameter monodisperse polystyrene latex with barium chloride concentration: 1. experimental values; 2. values calculated assuming additive conductivities and no interactions.

degree of dissociation α. Similarly, the dissociated H^+ ion concentration at different concentrations of potassium and barium chloride can be estimated potentiometrically (Figure 5), and the corresponding experimental values of apparent degree of dissociation can be determined. Table I gives the values of α determined conductometrically and potentiometrically along with the corresponding values for the ratios between the descending and ascending slopes of the conductometric titration curves. For the conductometric titration of hydrochloric acid with sodium hydroxide (strong

TABLE I

Apparent Degree of Dissociation from Conductometric and Potentiometric Titrations

234nm-diameter monodisperse polystyrene latex

Electrolyte		Conductometric Titration		Potentiometric Titration
Type	Concentration M, x10^5	Ratio of Slopes	α	α
none	----	0.136	0.0825	0.088
KCl	0.5	0.406	0.252	0.266
KCl	1.0	0.515	0.361	0.337
KCl	1.5	0.585	0.443	0.363
$BaCl_2$	0.25	0.466	0.337	0.473
$BaCl_2$	0.50	0.934	0.610	0.790
$BaCl_2$	0.75	1.116	0.775	0.920

acid with strong base), this ratio was found to be 1.215. For ion-exchanged polystyrene latexes, the values of this ratio were only 0.20-0.25 (2, 3), in agreement with the values found for acidified silver iodide sols (0.179 by de Bruyn and Overbeek (11) and 0.24 by van Os (12)).

Figure 6 compares the experimental and theoretical conductometric titration curves for the ion-exchanged 234nm-diameter polystyrene latex, assuming complete dissociation of the sulfate groups in H^+ or Na^+ forms. Experimental conductivities of the latex are smaller than the theoretical conductivities throughout the neutralization process (descending leg of the curve), which suggests that the surface sulfate groups are not completely dissociated in either the Na^+ or H^+ form.

This observation led to the development of a new experimental approach to investigate the apparent degree of

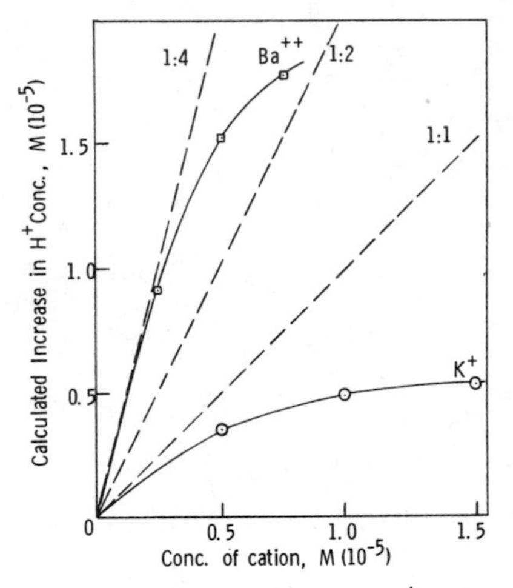

Figure 5. Variation of the increase in H^+ ion concentration (from pH measurement) with added cation concentration for ion-exchanged 234nm-diameter monodisperse polystyrene latex.

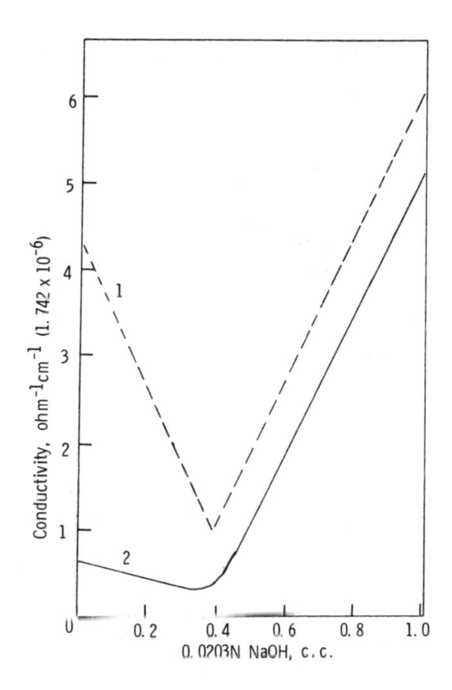

Figure 6. Conductometric titration of ion-exchanged 234nm-diameter monodisperse polystyrene latex: 1. theoretical curve calculated assuming 100% dissociation; 2. experimental curve.

dissociation of the sulfate groups in different cation forms (i.e., H^+, Na^+, Ba^{++}). Furthermore, the results of this investigation suggest that interfacial ion exchange occurs whenever electrolyte is added to the ion-exchanged latex. The addition of electrolyte results in a sudden increase in the ion concentration in the aqueous phase, which upsets the equilibrium state existing between the ions in the surface and those in the aqueous phase; therefore, ion-exchange takes place between the ions in the aqueous phase and those in the surface phase until a new equilibrium state is reached. The equilibrium state existing between the ions present in the surface phase and those in the aqueous phase of the latex is a function, not only of the surface characteristics (especially the surface charged groups) of the latex particles, but also the nature of the ions (cations and anions) present. The final ionic equilibrium state is determined by the attractive interactions (between the cations and the negatively-charged surface groups, the anions and the positively-charged groups, the cations and the anions, etc.) and the repulsive interactions (between the cations and the positively-charged surface groups, the anions and the negatively-charged surface groups, the cations and the cations, the anions and the anions, etc.) between the ions present and the particle surface groups.

For an ion-exchanged polystyrene latex with only surface sulfate groups in the H^+ form, the only ions present in significant quantity are the H^+ ions. Such a latex is an ideal model system to study the interaction between the sulfate groups of the latex particles and the H^+ ions. A 357nm-diameter monodisperse polystyrene latex with only surface sulfate groups was selected for studying the interaction of this type. The latex was first cleaned thoroughly by dialysis to remove the bulk of the electrolyte, followed by ion exchange to remove the remaining electrolyte and convert the surface sulfate groups to the H^+ form. The concentration of the sulfate groups of the latex was then determined quantitatively by the conductometric titration method described earlier (11) and was found to be 5.5×10^{-6}gm eq/gm solids.

Figure 7 shows the variation of conductivity and H^+ ion concentration of this latex with solids content; Line 1 shows the theoretical variation of the H^+ ion concentration assuming 100% dissociation of the sulfate endgroups; Line 2 shows the theoretical variation of conductivity; Line 3 shows experimental variation of H^+ ion concentration from pH measurement; Line 4 shows the expected variation of

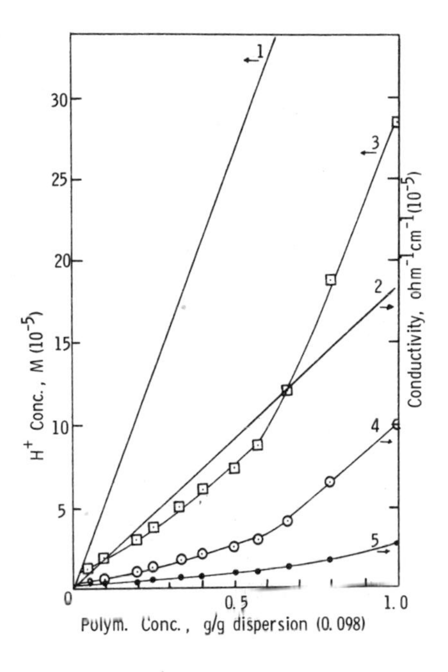

Figure 7. Variation of H^+ ion concentration and conductivity with polymer concentration for ion-exchanged 357nm-diameter monodisperse polystyrene latex: 1. theoretical H^+ ion concentration; 2. theoretical conductivity; 3. experimental H^+ ion concentration; 4. theoretical conductivity based on experimental pH measurement; 5. experimental conductivity.

conductivity based on experimental pH measurements; Line 5 shows the experimental variation of conductivity.

Table II gives the electrophoretic mobilities of the cleaned 357nm-diameter monodisperse polystyrene latex in the H^+, Na^+ and Ba^{++} forms and the corresponding values of the apparent degree of dissociation of the sulfate end-

TABLE II

Electrophoretic Mobility and Degree of Dissociation

Results for 357nm-diameter Polystyrene Latex

(0.35% solids)

Counter-Ion	Electrophoretic Mobility, U_e, cm^2/volt-sec.	Degree of Dissociation, α
H^+	1.41×10^{-4}	0.063
Na+	5.12×10^{-4}	0.70
Ba^{++}	1.52×10^{-4}	0.10

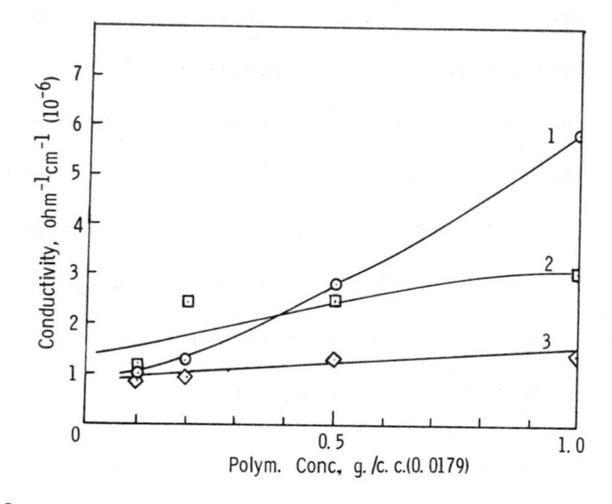

Figure 8. Variation of conductivity with polymer concentration of ion-exchanged monodisperse polystyrene latex in different cation forms: 1. H^+; 2. Na^+; 3. Ba^{++}.

groups determined from the experimental electrophoretic mobility, conductivity, and conductometric titration results. To eliminate any possible ion-exchange effect in the study of electrolyte effect on the electrophoretic mobilities of the latex particles, only the electrolyte with the same cation as the counterion of the cleaned latex was added.

Preliminary experiments were carried out to determine the effect of solids content and electrolyte concentration on the electrophoretic mobilities of the cleaned latex. Figure 8 shows the variation of the conductivities of the cleaned 357nm-diameter monodisperse polystyrene latex in the H^+, Na^+, and Ba^{++} forms with solids content. Figure 9 shows the variation of the electrophoretic mobilities of the same latex in the H^+, Na^+, and Ba^{++} forms with solids content. Figure 10 shows the variation of the electrophoretic mobilities of the same latex with the corresponding electrolyte concentration. The corresponding electrolytes for the latex in the H^+, Na^+, and Ba^{++} forms are HCl, NaCl, and $BaCl_2$, respectively, e.g., for the cleaned latex in the H^+ form, only the effect of hydrochloric acid concentration was studied.

DISCUSSION

The structure of a cleaned latex particle with sulfate endgroups in the H^+ form can be represented in terms of the electric double layer. Let $\sigma_{SO_4^-,S}$, $\sigma_{H^+,\delta}$, and $\sigma_{H^+,G}$ represent the charge densities at the surface of the particle, the Stern layer, and the Gouy layer, respectively, to give the following relationship

$$\sigma_{SO_4^-,S} = \sigma_{H^+,\delta} + \sigma_{H^+,G} \qquad (9)$$

assuming that the H^+ ions are the only counterions present in the latex (which is a valid assumption because

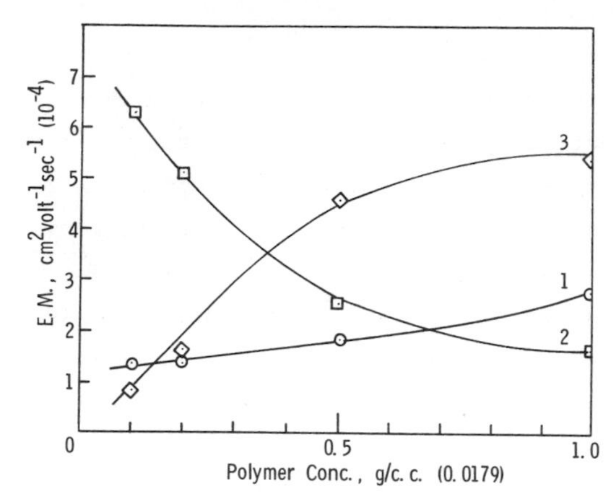

Figure 9. Variation of electrophoretic mobility with polymer concentration of ion-exchanged 357nm-diameter monodisperse polystyrene latex in different cation forms: 1. H^+; 2. Na^+; 3. Ba^{++}.

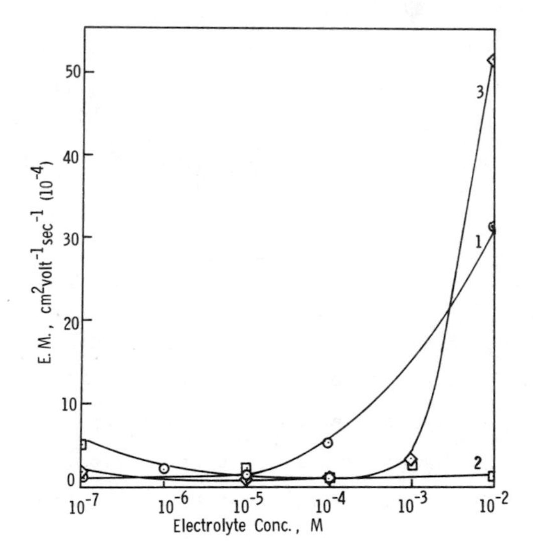

Figure 10. Variation of electrophoretic mobility with log electrolyte concentration of ion-exchanged 357nm-diameter monodisperse polystyrene latex in different cation forms: 1. H^+ in hydrochloric acid; 2. Na^+ in sodium chloride; 3. Ba^{++} in barium chloride.

the conductivities of the serum water obtained from ion-exchanged polystyrene latexes were found to be 0.6-0.8 μmhos, same value as for deionized water). The surface charge density $\sigma_{SO_4^-,S}$ can be determined from the conductometric titrations, but there are no direct ways to determine the values of $\sigma_{H^+,\delta}$ and $\sigma_{H^+,G}$. Nevertheless, qualitatively, the picture is clear: $\sigma_{SO_4^-,S}$ is independent of any change in the aqueous phase, while any increase in the value of $\sigma_{H^+,\delta}$ in the Stern layer would decrease the Stern potential Ψ_δ and the value of $\sigma_{H^+,G}$ in the Gouy layer. Therefore, the addition of H^+ ions to the latex would cause some H^+ ions to migrate into the Stern layer (adsorb), resulting in a decrease in Ψ_δ and the stability of the latex. However, if an electrolyte with a different counterion (e.g., Na^+) were added to the clean latex in H^+ form, the relationship would become more complex. This point will be discussed later.

First, however, let us consider the cleaned polystyrene latex with other cations such as Na^+ or Ba^{++} as the only counterions in the latex. The value of $\sigma_{SO_4^-,S}$ would remain the same, but the values of $\sigma_{Na^+,\delta}$ and $\sigma_{Ba^{++},\delta}$ and the corresponding Stern potentials, and the values of $\sigma_{Na^+,G}$ and $\sigma_{Ba^{++},G}$ would vary according to the adsorption behavior of the respective counterions.

Figure 7 shows that, in the cleaned 357nm-diameter polystyrene latex, there are H^+ ions of three types: 1. H^+ ions adsorbed in the Stern layer and in the diffuse Gouy layer which cannot be detected by pH or conductivity measurements; 2. H^+ ions in the diffuse layer which can be detected by pH measurement, but not by conductivity measurement; 3. H^+ ions which can be detected by conductivity and pH measurements.

The structure of a cleaned polystyrene latex particle in the H^+ form in normal neutral environment would have the counterions distributed symmetrically around the particle. However, when the particle is subjected to an electric field of strength $\underline{X}$ (the actual condition in which the conductivity measurements are carried out), the particles migrate toward the electrode of opposite charge, with the ionic field lagging behind according to the particle size and electrolyte concentration (relaxation effect).

The measured conductivity λ $(ohm^{-1}cm^{-1})$ of the cleaned latex can be given as follows

$$\lambda = \lambda_{particle} + \lambda_{H^+} + {}^{o}\lambda_S \qquad (10)$$

$$= NqU_e + N_{H^+}eU_{H^+} + {}^{o}\lambda_S \qquad (10a)$$

where $\lambda_{particle}$ is the conductivity due to the latex particles, λ_{H^+} the conductivity due to the H^+ ions present, ${}^{o}\lambda_S$ the conductivity of the serum, $\underline{N}$ the number of particles/cc latex, $\underline{q}$ the net charge carried by each particle in the electric field, $\underline{U}_e$ the electrophoretic mobility of the particles (cm^2/volt sec), $\underline{N}_{H^+}$ the number of H^+ ions/cc latex, $\underline{e}$ the elementary charge, and $\underline{U}_{H^+}$ the mobility of H^+ ions in water at 25°C (cm^2/volt sec). For electrical neutrality of the system

$$Nq = N_{H^+} e \qquad (11)$$

From Equation 10a

$$\lambda = Nq(U_e + U_{H^+}) + {}^{o}\lambda_S \qquad (12)$$

All the terms in Equation 12 except q can be measured experimentally, so that substituting in this equation gives the value of q. The apparent degree of dissociation, defined as the ratio of the experimentally-determined charge per particle $\underline{q}$ to the theoretical surface charge determined by conductometric titration can then be determined. Similarly, for the cleaned polystyrene latex with other cations as the only counterions,

$$\lambda = Nq(U_e + U_i) + {}^{o}\lambda_S \tag{13}$$

where U_i is the mobility of the ion $\underline{i}$ in water at 25°C.

Table II gives experimental values for the electrophoretic mobilities of the 357nm-diameter polystyrene latex particles and the corresponding values of the apparent degree of dissociation α where the metal ions are the only counterions present in the cleaned latex. The cleaned latex with the Ba^{++} counterions showed the smallest value of α, while that with the Na^+ counterions showed the highest value.

Equation 13 can be modified to show the effect of electrolytes on the values of $\underline{q}$ and α. The significance of the parameters $\underline{q}$ and α is that they can be determined experimentally; $\underline{q}$ is a measure of the particle charge and α an experimental estimate of the distribution of the counterions when the particles are subjected to an electric field. Qualitatively, the higher the apparent degree of dissociation, the higher the corresponding values of q and Ψ_δ, and the more stable the latex.

Figure 2 shows the variation of the conductivity of the 234nm-diameter polystyrene latex in the H^+ form with potassium chloride concentration. The added K^+ ions may adsorb in the Stern layer, enter the diffuse Gouy layer, or remain in the bulk aqueous phase; the electrophoretic mobility and the corresponding net charge per particle may change as the apparent degree of dissociation of the sulfate endgroups varies according to the additional K^+ and Cl^- ions. The total effect is an increase in conductivity over and above that attributable to the added potassium chloride. This extra conductivity may be due to the exchange of the H^+ ions in the Stern and Gouy layers for the added K^+ ions or to the increase in the apparent degree of dissociation.

Figure 4 shows similar but much more pronounced effects when barium chloride was added to the 234nm-diameter polystyrene latex in the H^+ form.

Figure 5 shows that the ion-exchange phenomenon can also be studied potentiometrically. The initial addition of K^+ ions causes an increase in the H^+ ion concentration in the diffuse Gouy layer and the aqueous bulk phase, corresponding almost exactly to the theoretically-predicted value of one. These H^+ ions migrated from the Stern and Gouy layers where they could not be detected by the pH measurement. Further additions cause lesser increases. The results can be interpreted as either ion-exchange or an increase in the apparent degree of dissociation. The addition of barium chloride to the latex causes both ion-exchange and an increase in the apparent degree of dissociation. The increase in H^+ ion concentration as a function of added Ba^{++} ion concentration is much greater than the theoretically-predicted value of two. The extra H^+ ions present must be attributed to an increase in the apparent degree of dissociation.

The conductometric titrations of the cleaned 234nm-diameter polystyrene latex shown in Figures 1 and 3 can be interpreted as follows. When the cleaned latex with no added electrolyte is titrated with Na^+ and OH^- ions, the Na^+ ions may adsorb in the Stern layer, enter the diffuse Gouy layer and ion exchange with the H^+ ions present there, or remain in the bulk aqueous phase; the OH^- ions may neutralize the H^+ ions in the Stern layer or the diffuse Gouy layer. Consequently, there are two reactions taking place simultaneously, i.e., ion exchange and neutralization. The linearity of the descending leg of the conductometric titration curve indicates that the OH^- ions reacted indiscriminately with the H^+ ions present in the system (only those which are dissociated would contribute to the conductivity). The descending leg would be curved if the OH^- ions reacted preferentially with the H^+ ions in certain layers. As more and more Na^+ ions replace the H^+ ions as the counter ions during the titration, the apparent degree of dissociation of the surface sulfate endgroups may vary, but the change in conductivity due to this effect is small. A decrease in conductivity, however small, on adding sodium hydroxide to the latex means that the neutralization between the OH^+ and H^+ ions predominates over the ion-exchange between the Na^+ and the H^+ ions.

In the presence of electrolytes such as potassium or barium chloride, the titration is similar but more complex; some ion exchange has already taken place, and some H^+ ions have been released to the bulk aqueous phase. Upon adding Na^+ and OH^- ions to the system, neutralization between the OH^- and H^+ ions takes place in the aqueous phase, the Gouy layer, and the Stern layer, while very little ion exchange occurs between the H^+ ions in the surface layer and the Na^+ ions in the aqueous phase because the surface layers already contain substantial concentrations of K^+ or Ba^{++} ions. The result is a larger decrease in conductivity per unit volume of sodium hydroxide compared with the same cleaned latex without any added electrolyte.

If ion exchange predominates over neutralization during the conductometric titration of a cleaned latex

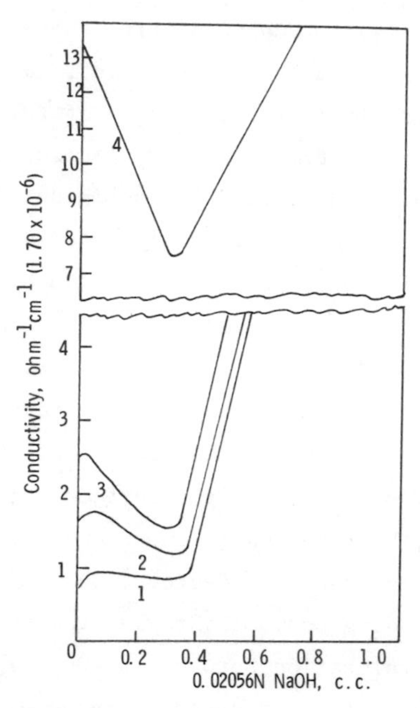

Figure 11. Conductometric titration of ion-exchanged 440nm-diameter monodisperse polystyrene latex as a function of sodium chloride concentration: 1. none; 2. 5 x 10^{-7}M; 3. 1 x 10^{-6}M; 4. 1 x 10^{-4}M.

with sodium hydroxide solution, the conductivity and H^+ ion concentration in the aqueous phase should increase upon addition of sodium hydroxide to the latex. Figure 11 shows such an example, the conductometric titration of a cleaned 440nm-diameter monodisperse polystyrene latex. With no added electrolyte, the ion exchange process initially predominated over the neutralization, resulting in an increase in conductivity and

H^+ ion concentration (the pH actually decreased). As more sodium hydroxide solution was added, the neutralization process became predominant, and the secondary maximum shown in Figure 13 represents the critical point where there was no change in conductivity when sodium hydroxide was added to the latex (i.e., the conductivity decrease due to neutralization between the OH^- and H^+ ions was nullified by the conductivity increase due to ion exchange between the Na^+ and H^+ ions). The addition of small amounts of sodium chloride to the cleaned polystyrene latex before titration resulted in the elimination of the first ascending part of the curve, as shown by lines 2, 3, and 4 of Figure 11, indicating the predominance of the neutralization process over the ion exchange process because of the saturation of the inner surface with Na^+ ions of the electrolyte.

Thus, the conductometric titration (Figures 1 and 3), the conductometric (Figures 2 and 4), and the potentiometric (Figure 5) results, and the corresponding calculated values of the apparent degree of dissociation of the sulfate endgroups of the polystyrene latex as a function of electrolyte (KCl or $BaCl_2$) concentration, (Table 1) may be interpreted qualitatively in terms of both ion exchange and neutralization processes taking place concurrently between all of the ions present in the latex system.

From the variation of conductivity with solids content for the 357nm-diameter cleaned latex (Figure 8), the conductivities of the cleaned latex in the H^+ and Na^+ forms did not increase linearly with increasing solids content. For the latex in the H^+ form, the curve deviated positively from linearity, (increased in slope) indicating an increase in the apparent degree of dissociation α with increasing solids content. For the latex in the Na^+ form, the curve deviated negatively from linearity, (decreased in slope) indicating a decrease in α with increasing solids content. For the latex in the Ba^{++} form, there was no significant deviation from linearity; the value of α was independent of solids content.

The variation of electrophoretic mobility with solids content for the same latex is shown in Figure 9. For the latex in the H^+ form, the electrophoretic mobility increased slightly with increasing solids content, while, for the same latex in the Na^+ form, the electrophoretic mobility decreased strongly with increasing solids content. The slight increase for the latex in the H^+ form parallels the change in the apparent degree of dissociation of the sulfate endgroups as indicated by the conductivity measurements for this latex. However, for the latexes in the Na^+ and Ba^{++} forms, the variation is more complex and other factors have to be considered to make any quantitative interpretation.

Figure 10 shows the variation of electrophoretic mobility with log electrolyte concentration; the latex in the H^+ form used hydrochloric acid as electrolyte, that in the Na^+ form sodium chloride, and that in the Ba^{++} form barium chloride, to obviate ion exchange with the cation of the added electrolyte. The variation of electrophoretic mobility correlates approximately with the apparent degree of dissociation determined for the ion-exchanged latexes in the H^+, Na^+, and Ba^{++} forms, i.e., the electrophoretic mobility of the latex in the Na^+ form (high value of α) varied only slightly over a five-decade range of electrolyte concentration, while those in the H^+ and Ba^{++} forms (low values of α) showed a strong increase at a critical electrolyte concentration (10^{-5}M for the Na^+ form and 10^{-3}M for the Ba^{++} form).

In summary, Figures 8-10 show that the electrophoretic mobility depends upon the solids content, the nature of the counterions of the sulfate endgroups, and the type and concentration of electrolyte. Further theoretical and experimental studies of the effect of electrolyte on the electrophoretic mobility of the latex particles and the corresponding apparent degree of dissociation of the surface sulfate groups are needed to understand these effects more thoroughly.

CONCLUSIONS

1. The values of the apparent degree of dissociation α of the sulfate endgroups of ion-exchanged polystyrene latex particles can be determined experimentally using conductometric and potentiometric techniques.
2. The values of α vary with the type of counterions present in the cleaned latex.
3. Ion-exchange and neutralization take place concurrently when the ion-exchanged latex in the H^+ form is titrated with sodium hydroxide; the combination of these two phenomena determine the shape of the conductometric titration curve.
4. The apparent degree of dissociation effect may be interpreted qualitatively by the ion-exchange process.

REFERENCES

1. H. J. van den Hul and J. W. Vanderhoff, J. Colloid Interface Sci. 28: 336 (1968).
2. J. W. Vanderhoff, H. J. van den Hul, R. J. M. Tausk, and J. Th. G. Overbeek, "Clean Surfaces: Their Pre-

paration and Characterization for Interfacial Studies," G. Goldfinger, editor, Marcel Dekker, New York, 1970, p. 15.

3. H. J. van den Hul and J. W. Vanderhoff, British Polymer J. 2: 121 (1970).
4. H. J. van den Hul and J. W. Vanderhoff, "Polymer Colloids," R. M. Fitch, editor, Plenum Press, New York, 1971, p. 1.
5. G. D. McCann, E. B. Bradford, H. J. van den Hul, and J. W. Vanderhoff, "Polymer Colloids," R. M. Fitch, editor, Plenum Press, New York, p. 29.
6. H. J. van den Hul and J. W. Vanderhoff, J. Electroanal. Interfacial Electrochem. 37: 161 (1972).
7. R. H. Ottewill and J. N. Shaw, Kolloid Z. Z. Polymere 215: 161 (1967).
8. J. B. Smitham, D. V. Gibson, and D. H. Napper, J. Colloid Interface Sci. 45: 211 (1973).
9. G. D. McCann, J. W. Vanderhoff, A. Strickler, and T. I. Sachs, Separation Purification Methods 2: 153 (1973).
10. J. W. Vanderhoff, F. J. Micale, M. S. El-Aasser, and W. C. Wu, Polymer Preprints 16(1): 125 (1975).
11. W. C. Wu, Ph.D. Thesis, Lehigh University, 1977.
12. H. de Bruyn and J. Th. G. Overbeek, Kolloid Z. 84: 186 (1938).
13. G. A. J. Van Os, Ph.D. Thesis, University of Utrecht, 1943.

PREPARATION AND CHARACTERIZATION OF POLYSTYRENE LATEXES WITH THE IONIC COMONOMER 2-ACRYLAMIDO-2 METHYL PROPANE SULFONIC ACID

Richard L. Schild, Mohamed S. El-Aasser,
Gary W. Poehlein, and John W. Vanderhoff

Emulsion Polymers Institute
Lehigh University
Bethlehem, Pennsylvania

ABSTRACT

Two series of polystyrene latexes were prepared with the ionic comonomer 2-acrylamido-2-methyl propane sulfonic acid (AMPS) in the sodium and hydrogen ion forms. The effects of increasing the recipe concentration of functional comonomer on the coagulum level, particle size, surface charge density and latex stability were investigated.

Particle size was found to decrease and then to increase slightly with increasing concentration of AMPS. The coagulum level followed a similar trend. Characterization of cleaned latexes by conductometric titration showed strong acid groups on both latex series and carboxyl groups for the AMPS (H^+ form) where pH was not controlled. The surface charge density of strong acid groups was found to increase with AMPS concentration. Latex stability, as determined by coagulation kinetics, showed a general increase in stability with increasing AMPS concentration. Deviations in expected stability trends were attributed to the particle surface being a poor representation of the DLVO model due to possible superposition of steric and electrostatic stabilization effects.

INTRODUCTION

Recently, work has been done to study the effect of functional comonomers (also referred to as comonomeric emulsifiers or ionic comonomers) on the kinetics of emulsion polymerization and the properties of the latexes prepared (1-4). Because the ionic comonomer contains a functional group and a polymerizable double bond, it can be used to incorporate a fixed charge on the particle surface. Colloidal properties, such as latex stability, can then be investigated on well-characterized latexes from which particle size and surface charge can be controlled by the quantity of functional monomer charged to the system.

The use of ionic comonomers in emulsion polymerization in place of traditional surface active agents could offer several advantages. Since the ionic comonomer copolymerizes into the polymer matrix, preferably at the particle surface, it provides a source of fixed charge. Consequently, the physical properties of the latex system cannot be altered through an adsorption/desorption cycle as with traditional surfactants. Since the surface charge, and therefore stability, is fixed and constant, a smaller amount of comonomer is usually required to provide the same stability as an adsorbed surface active agent. The ionic comonomer is of low foaming nature due to its short chain length. Greater quantities of comonomer can then be added to achieve high levels of stability without foaming problems.

In the present investigation, two polystyrene latex series were prepared with increasing amounts of 2-acrylamido-2-methyl propane sulfonic acid (AMPS) to study the effect of AMPS concentration on particle size, the degree of incorporation of AMPS to the particle surface, the surface charge density, and the latex stability against electrolytes as a function of the pH of the latex system.

EXPERIMENTAL

I. Materials

Doubly distilled, deionized and deoxygenated water was used in all polymerizations. Styrene monomer was washed with sodium hydroxide solution and then distilled under a nitrogen atmosphere with reduced pressure to remove inhibitors and impurities. The ionic comonomer AMPS was of special process reagent grade and used as received from the Lubrizol Corporation. The structure of AMPS is

$$CH_2 = CH - \overset{\overset{\displaystyle O}{\|}}{C} - \overset{\overset{\displaystyle H}{|}}{N} - \underset{\underset{\displaystyle CH_3}{|}}{\overset{\overset{\displaystyle CH_3}{|}}{C}} - CH_2 - SO_3^- \; H^+$$

The AMPS was used in the H^+ form (AMPS (H^+)) for the first series of latexes and in the Na^+ form (AMPS (Na^+)) for the second latex series. The AMPS (Na^+) was prepared by making a 0.32M AMPS (H+) solution and titrating with base until a pH of 7 was reached. All other reagents used were of analytical grade and used as received.

The ion exchange resins, Dowex 50W-X4 (H^+) and Dowex 1-X4 (OH^-) were supplied by the Dow Chemical Corporation and were rigorously purified by the method developed by Vanderhoff et al (5).

II. Latex Preparation

Two series of polystyrene latexes were prepared by emulsion polymerization in 12 oz. bottles using traditional techniques. The bottles were charged with styrene monomer, AMPS comonomer, methanol, buffer, and water in the quantities shown in Table I. Methanol was used to increase the styrene solubility in the aqueous phase and encourage early copolymerization of styrene and AMPS leading to high incorporation of the latter to the particle surface.

The bottles were preheated for approximately 30 minutes at the reaction temperature of 70°C, removed for the injection of initiator solution, replaced and allowed to polymerize for 24 hours. The

TABLE I

Latex Recipes

All recipes contain:		
	styrene	40 g
	water	140 g
	methanol	20 g
	sodium bicarbonate	0.4 g
	potassium persulfate	0.4 g

Latex Code	[AMPS, Na^+] mol/l_{aq}	Latex Code	[AMPS, H^+] mol/l_{aq}
SA-27	0.001	SA-10	0.0
SA-28	0.010	SA-2	0.0121
SA-29	0.025	SA-1	0.0422
SA-30	0.050	SA-13	0.0724
SA-31	0.075	SA-14	0.151
SA-32	0.100		

bottles were tumbled end over end at 35 RPM in a constant temperature bath at 70° C.

At the end of the polymerization, the bottles were heated to 90° C for 2 hours to decompose the initiator. The latex was then filtered to remove all coagulum formed during the polymerization.

III. Particle Size

Particle size was determined by electron microscopy using a Philips 300 transmission electron microscope (TEM). One drop of highly distilled latex was placed on a copper grid and dried at room temperature. The grid was then platinum shadowed at an angle of 45°. Particle size was measured directly from the electron micrograph negatives using a calibrated magnifying eyepiece. Approximately 200 particles were counted for each latex.

The number average diameter was calculated from

$$\bar{D}_n = \frac{\sum n_i D_i}{n_i} \qquad (1)$$

The weight average diameter was calculated from

$$\bar{D}_w = \left(\frac{\sum n_i D_i^6}{\sum n_i D_i^3} \right)^{1/3} \qquad (2)$$

The latex dispersity is given by the uniformity ratio

$$U = \frac{\bar{D}_w}{\bar{D}_n} \qquad (3)$$

If the uniformity ratio is less than 1.05, the latex is considered to be monodisperse.

IV. Surface Characterization

A quantitative and qualitative analysis of the latex particle surface was performed by conductometric titration. The charge density and type of functional group were determined from the equivalent points on the titration curve.

The latexes were cleaned by repetitive batch ion exchanges with purified mixed bed (Dowex 1 and Dowex 50) resins. The ion exchange cycles were repeated with fresh resins until consistent results from the conductometric titrations were achieved.

The conductivity of each dilute latex ($\sim$2% solids) was monitored continuously while a 0.01875 N sodium hydroxide solution was added continuously. The amount and type of each surface functional group was determined from the equivalent points on the titration curve.

V. Ultracentrifugation

A preparative ultracentrifuge was used to separate the latex serum for microscopy studies to determine if polyelectrolyte formed during polymerization. Approximately 10 ml of latex was placed in a polycarbonate tube and then inserted into a swinging bucket rotor. The sample was spun at 35,000 RPM for thirty minutes. The serum was drawn off by syringe and saved for further centrifugations. The polymer particles were redispersed in water and centrifuged several times to encourage desorption of any polyelectrolyte adsorbed on the particle surface. The serum collected from each run was combined and centrifuged to remove any remaining latex particles. One drop of the serum was then diluted and placed on a carbon grid and dried for observation by the TEM.

VI. Coagulation Rate Studies

Latex stability can be determined by measuring the resistance of the latex to coagulation using an electrolyte. The rate at which coagulation occurs can be related directly to the stability factor W (6) by

$$W = \frac{K_R}{K_S} \tag{4}$$

where K_R = rate constant for rapid coagulation

K_S = rate constant for slow coagulation

The coagulation kinetics were measured by recording the change in optical density of the coagulating latex with a Cary 14 Spectrophotometer (Figure 1). Wu (7) related the rate of change of the optical density to the stability factor W by

$$W = \frac{K_R}{K_S} = \frac{\left[\frac{d\,(O.D.)}{dt}\right]_{t = o,\ C_E = c.c.c.}}{\left[\frac{d\,(O.D.)}{dt}\right]_{t = o,\ C_E}} \tag{5}$$

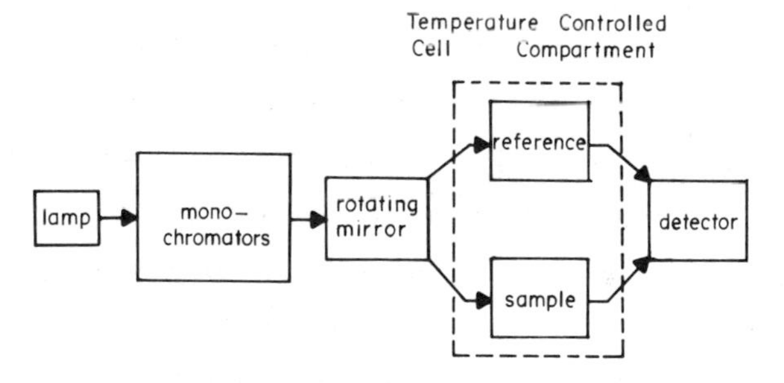

Figure 1. Schematic diagram of Cary 14 spectrophotometer used to measure optical density as a function of time to determine latex stability.

The rate constant or limiting slope for rapid coagulation is determined from the average slope of all rapid coagulation runs. Since in rapid coagulation, every collision is irreversible, all rapid coagulation cases will have approximately the same slope. Therefore, the limiting slope can be taken at the c.c.c. From the VO theory (7), it can be shown that W=1 for rapid coagulation.

The optical density versus time curves are determined by measuring the intensity ratio of the latex in the reference and sample cells. The quartz in sample cells have a 1 cm path length and 5 ml capacity. Three milliliters of dilute latex (0.020 wt %) was added to each cell. One ml of water was added to the reference cell. The cells were then placed into the spectrophotometer cell compartment where they were allowed to come to thermal equilibrium. The cell compartment was maintained at 25.0 $\pm$ 0.5°C by a constant temperature bath. One ml of $BaCl_2$ electrolyte solution of the appropriate concentration was injected into the sample cell by syringe. Optical density versus time was recorded immediately after injection. All runs were carried out at a wavelength of 1370 nm.

The stability ratio was determined by calculating the limiting slope from all rapid coagulation runs (W=1 for fast coagulation). Using the initial slopes of the slow coagulation runs, W values for each electrolyte concentration were calculated from equation 5. Log W versus Log C_E (electrolyte concentration) curves were then plotted as suggested by Reerink and Overbeek (8).

The c.c.c. (critical coagulation concentration) is defined as the point at which rapid coagulation changes to slow coagulation. It is a measurement of the latex resistance to coagulation. The c.c.c. can be used to calculate values for the Hamaker constant and the Stern potential. The Hamaker constant A is given by (3)

$$A = \left[\frac{1.73 \times 10^{-36} \left(\frac{d \text{ Log } W}{d \text{ Log } C_E} \right)^2}{a^2 \; v^2 \; (c.c.c.)} \right]^{1/2} \qquad (6)$$

where a = particle radius, cm
v = electrolyte valence
c.c.c. = critical coagulation concentration, moles/l

$\frac{d \text{ Log } W}{d \text{ Log } C_E}$ = slope of stability curve

C_E = electrolyte concentration, millimoles/l

This equation applies when $\kappa a \gg 1$ where κ = reciprocal double layer thickness (high electrolyte concentrations) and the flat plate model approximates that of a spherical system.

The slope of the stability curve can be related to the Stern Potential (8) by

$$-\frac{d \text{ Log } W}{d \text{ Log } C_E} = \frac{2.15 \times 10^7 \; a \; \gamma^2}{v^2} \qquad (7)$$

where $$\gamma = \frac{\exp\left(\frac{Ve\,\Psi_\delta}{2kT}\right) - 1}{\exp\left(\frac{Ve\,\Psi_\delta}{2kT}\right) + 1}$$

Ψ_δ = Stern potential, mv
e = electronic charge
k = Boltzman constant
t = temperature, °K.

RESULTS AND DISCUSSION

I. Coagulum

The coagulum levels for the latex series prepared with AMPS (Na^+) were recorded. Figure 2 shows the percent coagulum based on the monomer phase versus the AMPS (Na^+) concentration. The decreasing coagulum level with increasing AMPS (Na^+) concentration indicates that all primary particles formed are more stable due to greater incorporation of the AMPS (Na^+). At increased AMPS (Na^+) levels, excess functional monomer which does not copolymerize with styrene homopolymerizes to form polyelectrolyte. The polyelec-

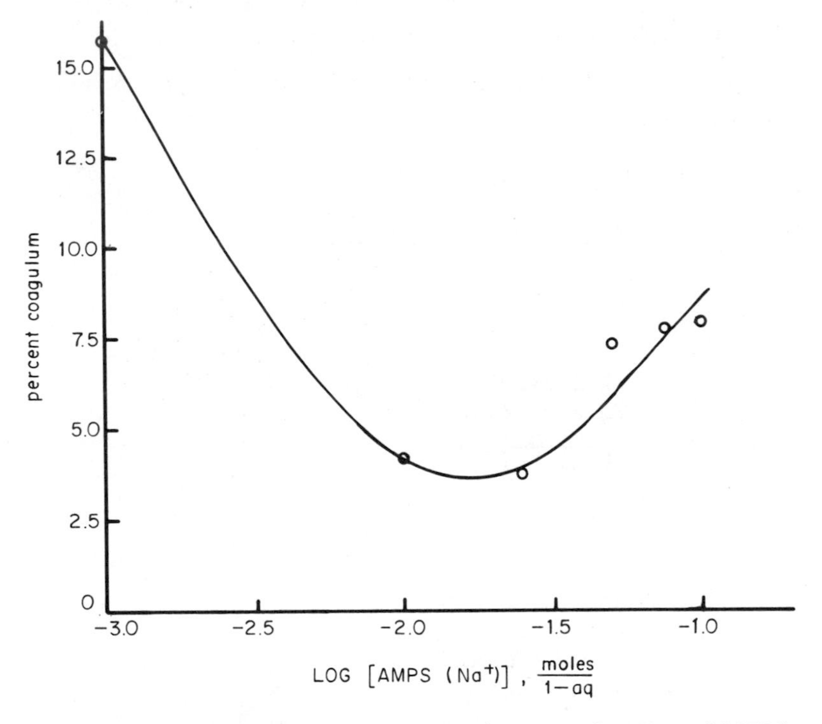

Figure 2. Variation of percent coagulum as a function of AMPS (Na^+) concentration.

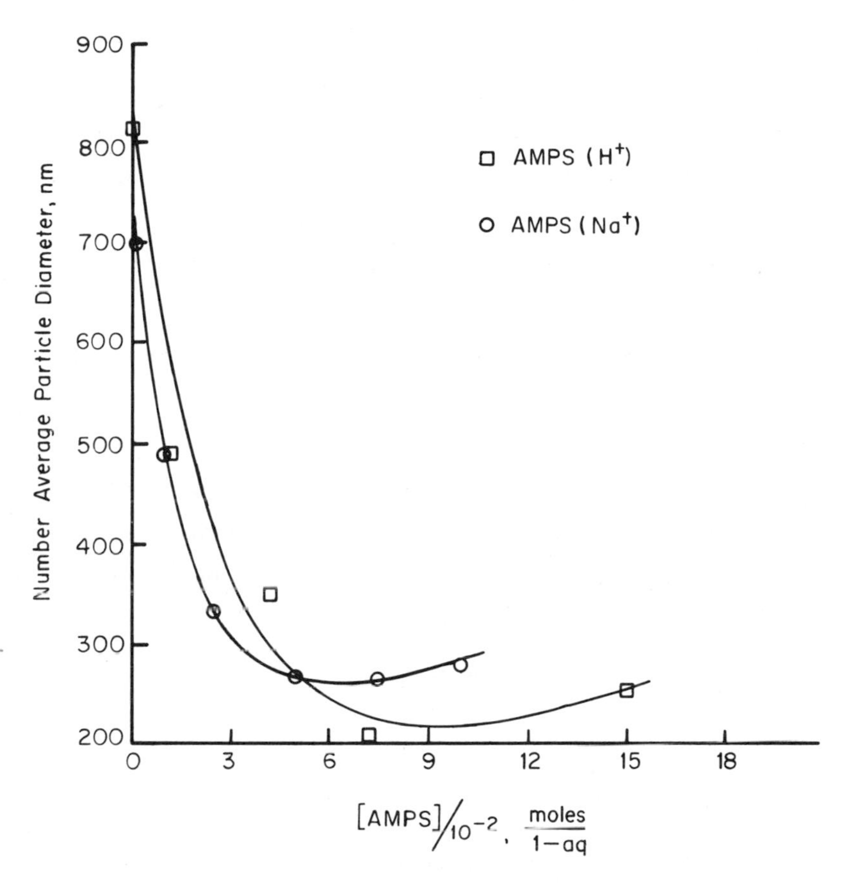

Figure 3. Variation of number average particle diameter as a function of AMPS concentration.

trolyte causes particle flocculation and causes the coagulum level and particle size to increase as evidenced in Figures 2 and 3.

II. Particle Size

The particle size of the latexes was determined by transmission electron microscopy. Figure 3 shows the number average particle diameter calculated from the photomicrographs versus the charged AMPS concentration. As the charged concentration of AMPS is increased, the particle size decreased to a minimum level. This observation can be explained by the following mechanisms. First, as the AMPS level increases, more of AMPS is incorporated into the primary particles formed making them more stable and less likely to flocculate. The monomer is then distributed over a greater number of polymerization sites resulting in a smaller final particle size. It is possible that as the AMPS level increases beyond a certain level, enough polyelectrolyte is formed by homopolymerization of excess AMPS to cause flocculation to result in an increasing particle size.

The technique used to produce the latexes resulted in preparing fairly monodisperse latexes. Table II shows the uniformity ratio for the AMPS (Na^+) latexes. A uniformity ratio of less than 1.05 is an indication of a monodisperse system. Several of the latexes had uniformity ratios of 1.002 or less, indicating very monodisperse latexes. Figure 4 shows an electron micrograph of latex SA-28 which had a uniformity ratio of 1.001.

TABLE II

Latex Dispersity

Latex	D_n (nm)	D_w (nm)	U
SA-10	816	856	1.048
SA-27	697	710	1.018
SA-28	487	488	1.001
SA-29	333	334	1.002
SA-30	266	278	1.046
SA-31	265	265	1.004
SA-32	280	280	1.001

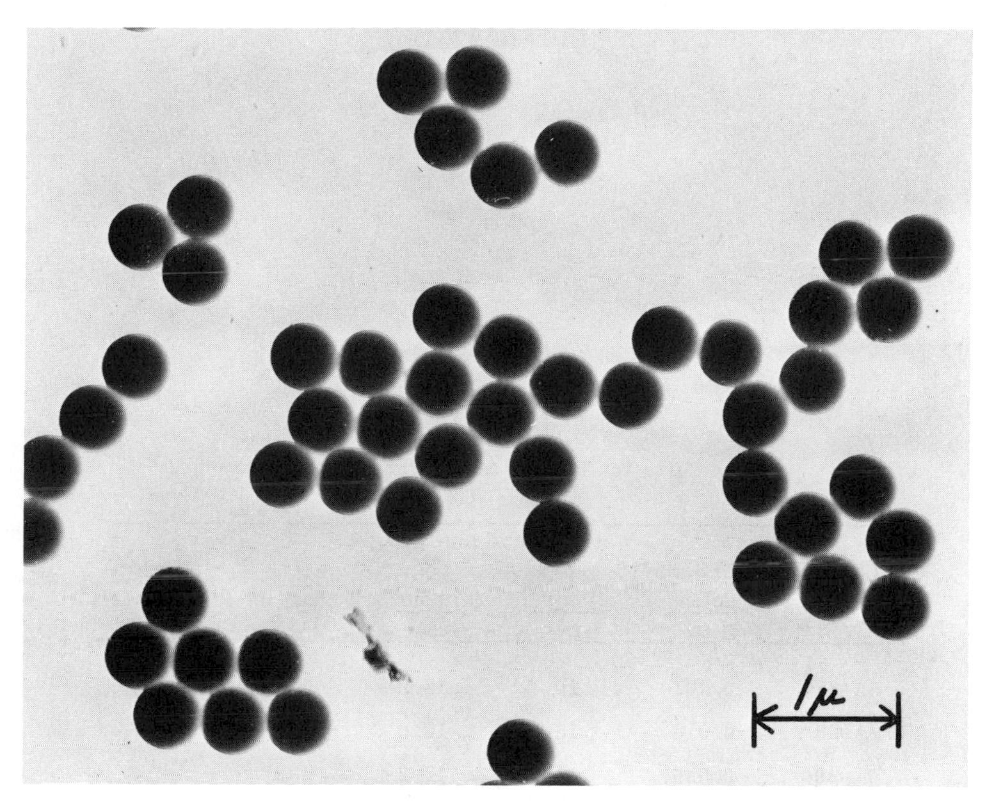

Figure 4. Electron micrograph of latex SA-28. $\bar{D}_n$ = 487 nm and U = 1.001

III. Conductometric Titration

After the latexes were cleaned by ion exchange with mixed bed resins, they were characterized by following the conductivity upon titration with 0.01875 N sodium hydroxide. The results of the titrations for the AMPS (Na^+) and AMPS (H^+) are shown in Tables III and IV respectively. Typical titration curves are shown in Figures 5 and 6 for AMPS in the Na^+ and H^+ form.

The pH of latexes measured after polymerization is also shown in Tables III and IV. The effect of pH on the presence of carboxyl groups is clearly evident for all latexes prepared with the AMPS

TABLE III

Titration Results for AMPS (Na^+) Series

Latex	[AMPS, Na^+] $\frac{moles}{liter}$	strong acid charges N_i $\frac{\mu eqv.}{gm}$	strong acid charges σ_s $\frac{\mu coul}{cm^2}$	$\frac{area}{charge}$ ($Å^2$)	pH after polymerization
SA-27	0.001	10.4	12.2	131	7.9
SA-28	0.010	15.7	12.9	124	7.7
SA-29	0.025	21.6	12.2	132	7.7
SA-30	0.050	32.3	15.0	111	8.2
SA-31	0.075	34.2	15.3	105	7.7
SA-32	0.100	40.7	19.3	83	7.5

TABLE IV

Titration Results for AMPS (H^+) Series

Latex	[AMPS, H^+] moles/liter	pH	Strong Acid Charges			Weak Acid Charges		
			N_i $\left(\frac{ueqv}{gm}\right)$	σ_s $\left(\frac{ucoul}{gm}\right)$	$\frac{area}{charge}$ (Å^2)	N_i $\left(\frac{ueqv}{gm}\right)$	σ_w $\left(\frac{ucoul}{gm}\right)$	$\frac{area}{charge}$ (Å^2)
SA-10	0.0	7.5	8.0	22.1	72.5	9.4	26.0	61.6
SA-2	0.0121	6.5	13.0	21.5	74.5	5.1	8.4	190.7
SA-1	0.0422	2.2	14.3	17.0	94.2	6.4	7.6	210.8
SA-13	0.0724	2.1	22.9	16.2	98.9	9.0	6.4	232.2
SA-14	0.151	6.8	45.0	38.7	41.4	19,9	17.1	93.7

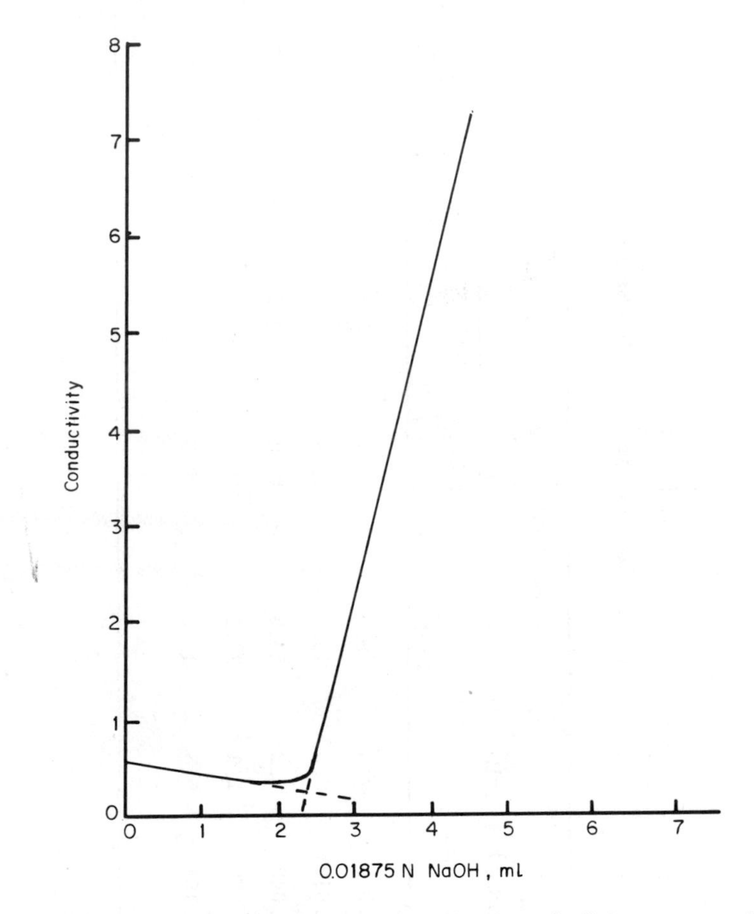

Figure 5. Typical conductometric titration of a latex prepared with AMPS (Na^+) comonomer.

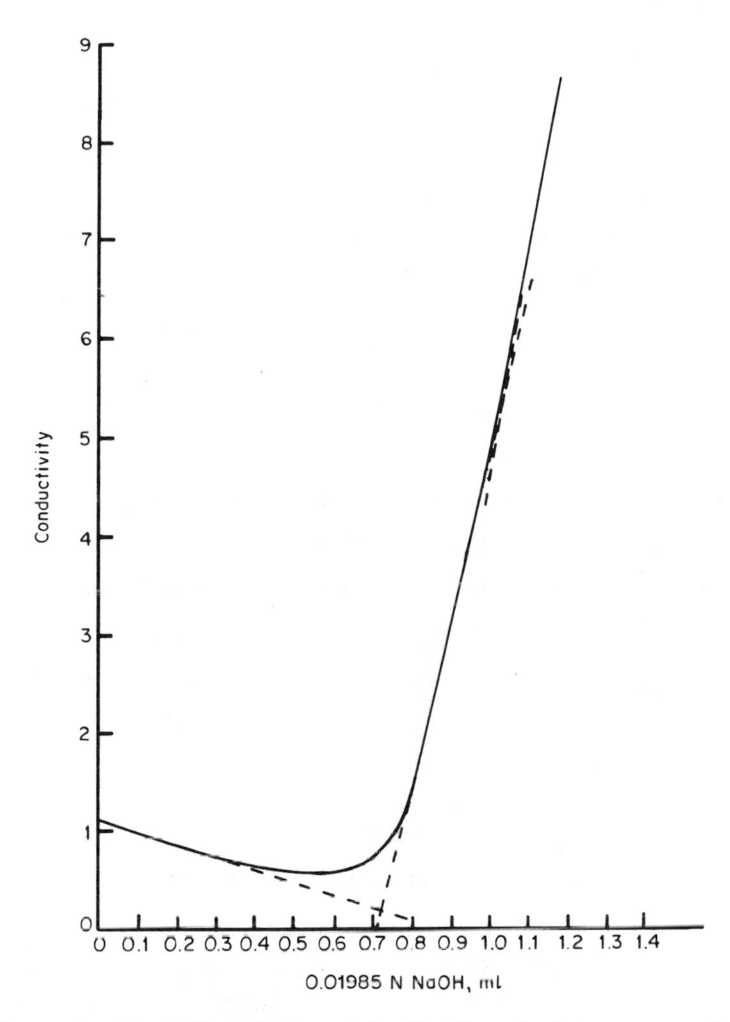

Figure 6. Typical conductometric titration of a latex prepared with AMPS (H^+) comonomer.

(H^+). The lower pH (less than 7) causes the hydrolysis of the sulfate initiator groups on the surface to form carboxyl groups. The amount of weak acid present seems to increase at higher concentrations of AMPS (H^+) where the pH is lower.

For all the latexes prepared, the total amount of strong acid charge increased with the increasing charge concentration of AMPS. The increase in the total amount implies that more of the AMPS has been incorporated into the particle surface. But by comparing the charge density of the AMPS (Na^+) to the total charge concentration, it is evident that the number of groups on the surface has only doubled while the AMPS charged concentration has increased 100-fold. Table V shows the amount of total charge that would be found on the surface due to the AMPS if total incorporation had occurred. As the charged concentration increased 100-fold (Latexes SA-27-32), the total amount of surface charge only increased four times. Therefore, the relative amount of the AMPS incorporated at the surface decreases with increased AMPS charge. A similar increase in strong acid charge was found for the latexes prepared with AMPS (H^+) (Table IV). The weak acid showed a varying charge density though decreasing to a minimum and then increasing with increasing AMPS concentration.

TABLE V

Incorporation of AMPS (Na^+) to Surface

Latex	[AMPS, Na^+] charged (mol/l $_{aq}$)	N_i (SO_4^- & SO_3^-) actual (μeqv/gm)	N_i (SO_3^-) expected from AMPS (μeqv/gm)
SA-27	0.001	10.4	3.5
SA-28	0.010	15.7	35.0
SA-29	0.025	21.6	87.5
SA-30	0.050	32.3	175.0
SA-31	0.075	34.2	263.0
SA-32	0.100	40.7	350.0

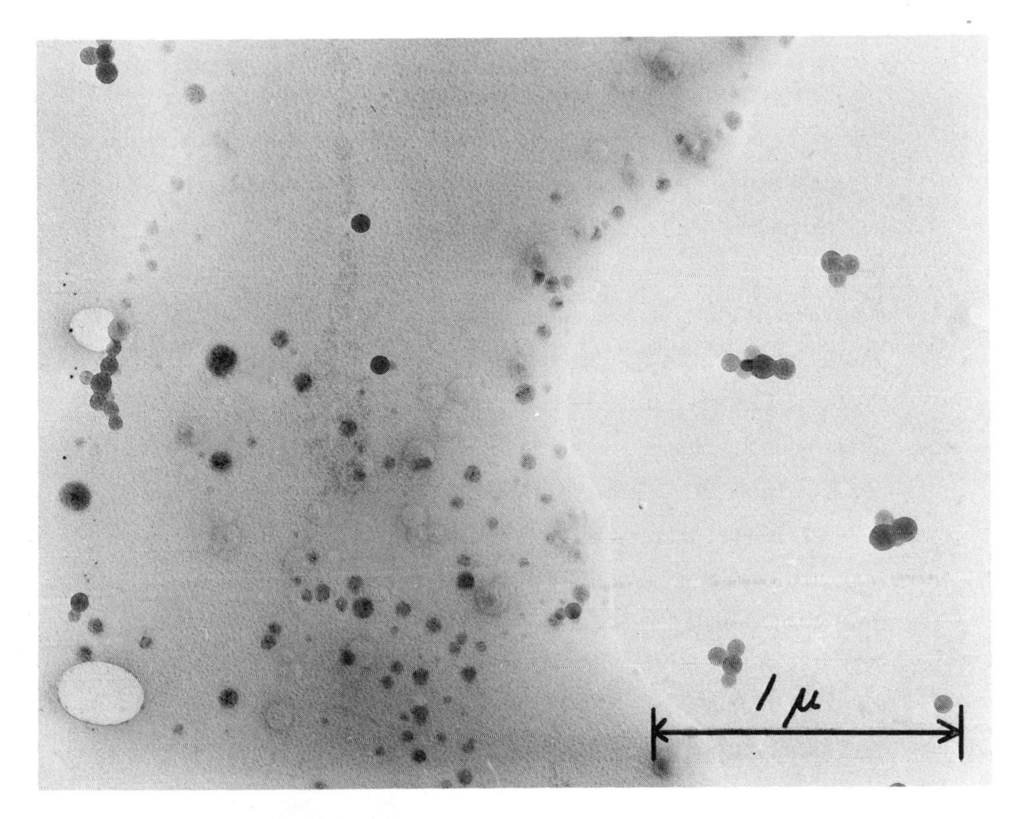

Figure 7. Electron micrograph of a polyelectrolyte film of AMPS isolated by ultracentrifugation of the serum phase of latex SA-13.

IV. Ultracentrifugation

The preparative ultracentrifuge was used to isolate the serum phase to detect the presence of polyelectrolyte. Figure 7 shows an electromicrograph of the serum phase of SA-13. A continuous film is shown which contains some latex particles. This film is probably due to polyelectrolyte formed by homopolymerization of AMPS. It is evident then that it is possible to form polyelectrolyte AMPS in emulsion polymerizations. This polyelectrolyte can cause particle flocculation thereby controlling particle size and coagulum levels.

V. Coagulation Rate Studies

The stability of ion-exchanged latexes was determined by following the coagulation rate as a function of electrolyte concentration. Figure 8 shows a series of coagulation runs at different electrolyte concentrations for Latex SA-30 where the O.D. was followed as a function of time. From these curves, the W values are calculated as a function of electrolyte concentration C_E as described earlier and plotted as in Figure 9. Figures 10 and 11 show the stability curves for the AMPS (Na^+) and AMPS (H^+) series respectively. The slopes of the stability curves and the critical coagulation concentrations are given in Table VI. For the AMPS (H^+) series, the latexes follow a general trend of increasing stability with charged concentration as indicated by the increase in c.c.c. and d Log W/d Log C_E values. Conversely, the AMPS (Na^+) series follows a more random trend.

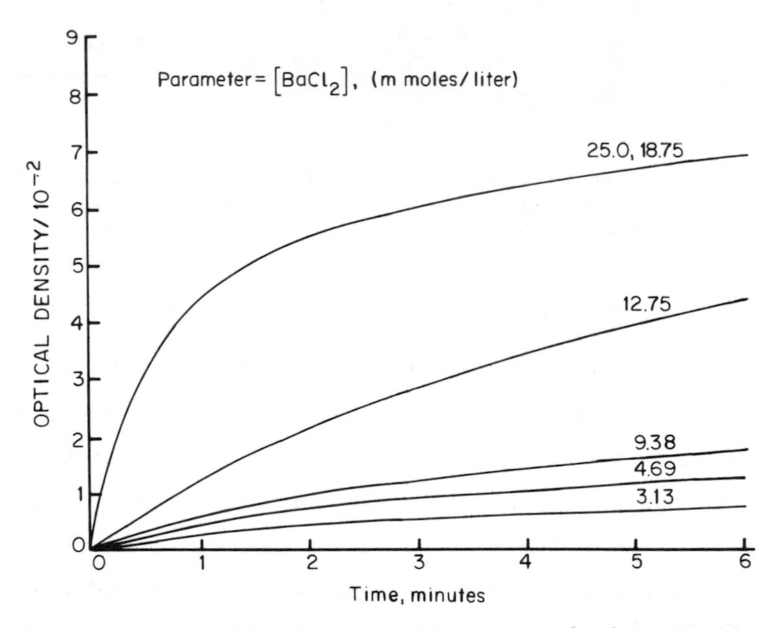

Figure 8. Optical density versus time curves for latex SA-30 used to determine coagulation rates. Wavelength of source = 1370 nm, $BaCl_2$ electrolyte.

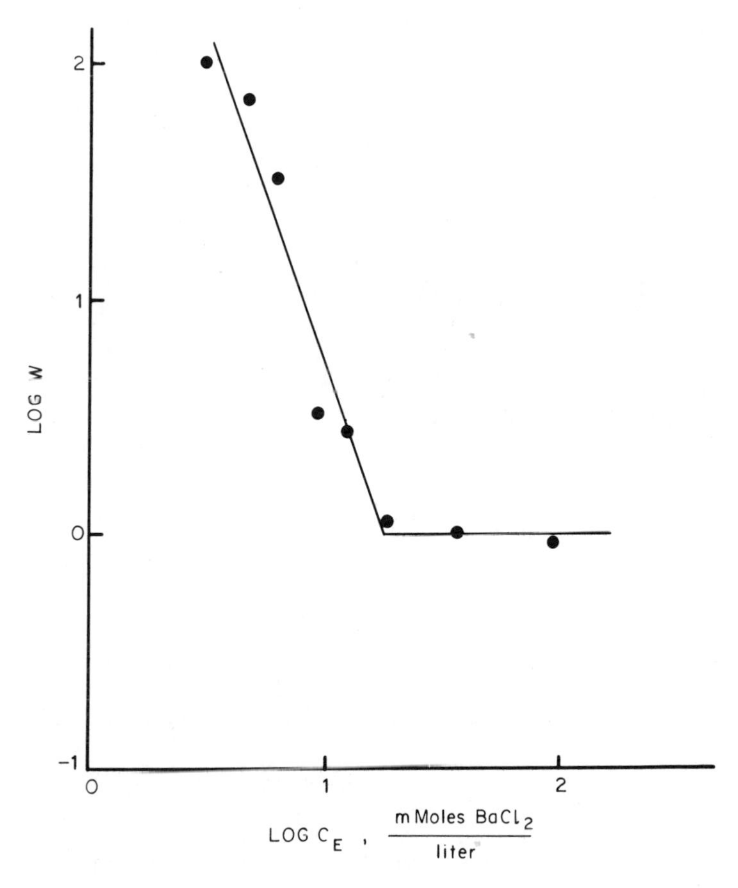

Figure 9. Stability curve for latex SA-32. c.c.c. = 18.3 mM $BaCl_2$.

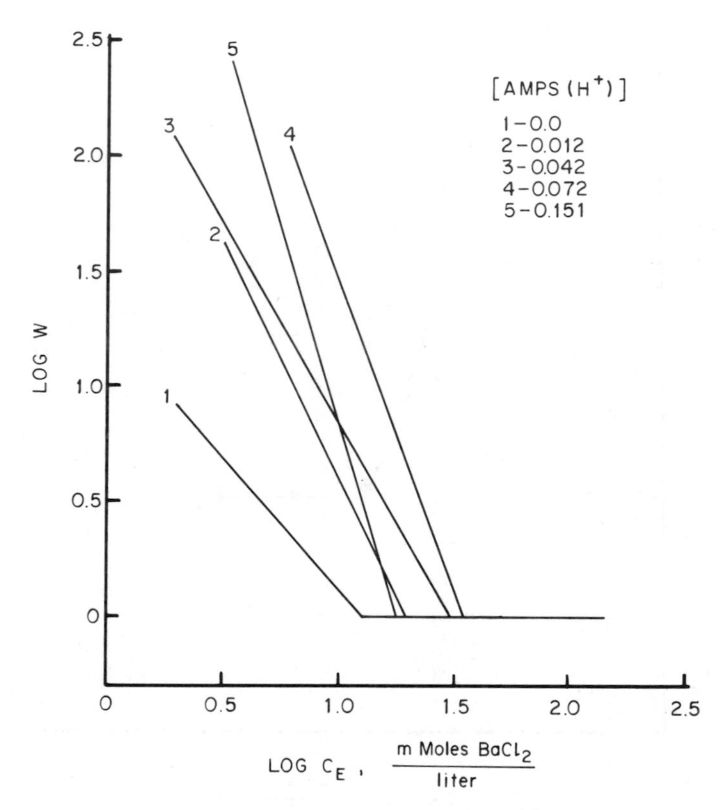

Figure 10. Stability curves for AMPS (H^+) latex series.

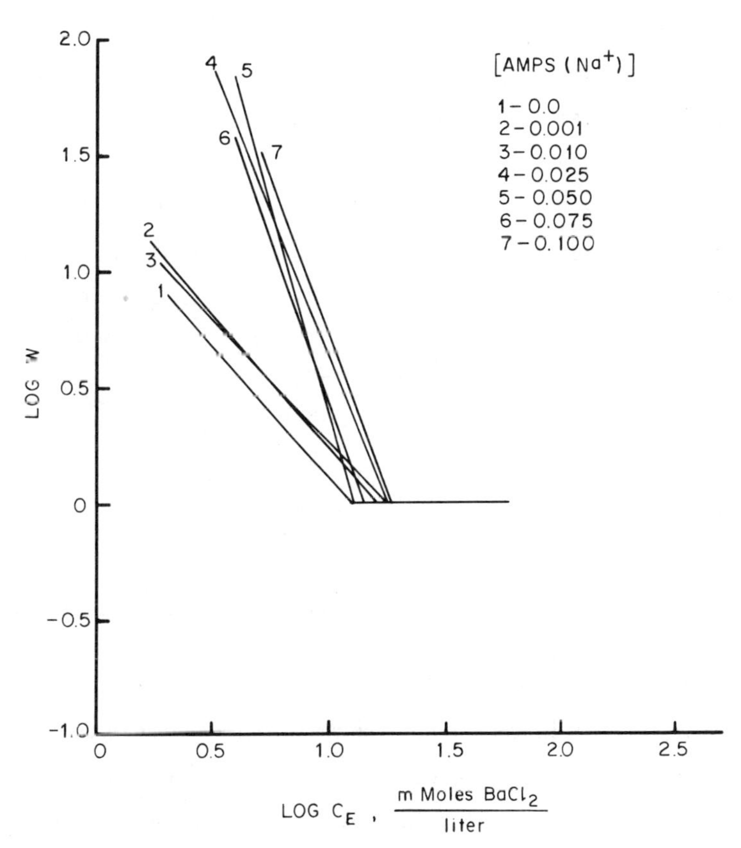

Figure 11. Stability curve for AMPS (Na^+) latex series.

TABLE VI

Results of Stability Analysis

Latex	$\bar{D}_N$ (nm)	σ_s (ueqv/gm)	σ_w (ueqv/gm)	$\frac{-d \text{ Log } W}{d \text{ Log } C_E}$	c.c.c. (m mole/l)
SA-27	697	10.4	--	1.15	16.8
SA-28	487	15.7	--	1.08	17.4
SA-29	333	21.6	--	2.57	18.0
SA-30	266	32.3	--	3.57	13.1
SA-31	265	34.2	--	2.88	14.5
SA-32	280	40.7	--	2.81	18.3
SA-10	816	22.1	26.0	1.19	12.5
SA-2	489	21.5	8.4	2.10	19.7
SA-1	351	17	7.6	1.76	30.8
SA-13	209	16.2	6.4	2.76	34.3
SA-14	254	38.7	17.1	3.42	17.7

The DLVO theory (7) predicts several possible trends for stability. First, as the surface charge density increases, the colloid stability will increase. Second, as particle size increases, the stability will increase. But this latter criterion assumes coagulation by the primary maximum. Wiese and Healy (9) showed that if flocculation occurred in the secondary minimum, the stability of the latex would decrease as particle size increased. Furthermore, secondary minimum flocculation would most likely occur in particles of diameter greater than 100 nm.

Table VI shows how the c. c. c. and stability curve slopes vary with AMPS (H^+ and Na^+) concentration. The stability analysis shows an increase in stability as the AMPS (H^+) concentration increases. This is expected in terms of the DLVO theory since the surface charge density increases and particle size decreases provided coagulation occurs in the secondary minimum. A similar, although more random trend, is observed for the AMPS (Na^+) series. The presence of unremoved polyelectrolyte and pennant chains of AMPS copolymerized to the surface could provide steric stability to the latex. The steric effect would be observed by increasing the electrolyte stability of a latex creating the random effects observed.

The steric contributions to the latex stability could be attributed to large functional groups chemically bound as well as polyelectrolyte strongly adsorbed to the particle surface. Figure 12 shows several possible configurations for copolymerization of AMPS to the surface and the location of charge. The AMPS can copolymerize in two ways exposing charge into the aqueous phase. First, a single AMPS comonomeric unit could copolymerize with styrene monomer, which would expose one charge group of the more hydrophilic AMPS molecule into the aqueous phase. The second possibility is the formation of a pennant chain. Here a persulfate-free radical has caused several AMPS molecules to form an AMPS oligomer which eventually picks up styrene molecules in the aqueous phase to form a primary particle. Such a pennant chain could extend far into the aqueous phase contributing both sterically and electrostatically to the latex stability. The individual AMPS molecules which extend from the pennant chain backbone would occupy a large amount of surface area. The pennant chain would therefore offer a large steric stabilizing effect. The concentrated charge on the pennant

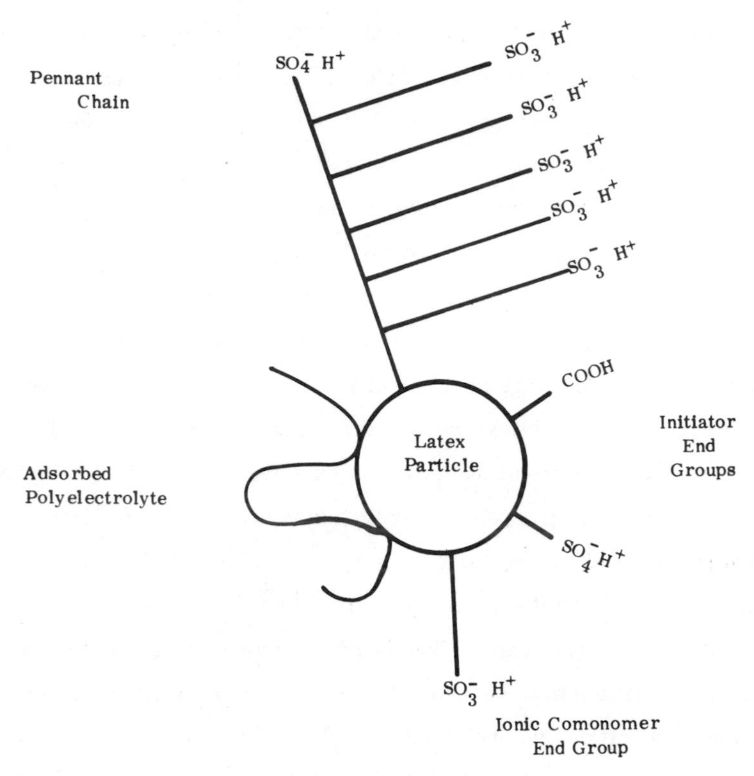

Figure 12. Conceptual representation of possible configurations of sources of stabilizing charge on the particle surface.

chain offers increased electrostatic effects due to the interactions of the ionic groups as well (10).

The Hamaker constant and Stern potential calculated from equations 6 and 7 are given in Table VII. The Stern potential increases with charged AMPS concentration for both series indicating increased stability. The Hamaker constant was found to be of the same order of magnitude as that calculated by Fowkes (11) for polystyrene (5×10^{-14} ergs) and experimentally determined by other workers (12, 13). The change in the value of the Hamaker constant

TABLE VII

Hamaker Constant & Stern Potential

Latex	A (ergs/10^{-14})	Ψ_δ (mvolts)
SA-27	0.53	4.0
SA-28	0.70	4.7
SA-29	2.39	8.8
SA-30	4.88	11.7
SA-31	3.75	10.5
SA-32	3.09	10.1
SA-10	0.54	3.8
SA-2	1.27	6.5
SA-1	1.19	7.1
SA-13	2.97	11.6
SA-14	4.21	11.7

as a function of the AMPS concentration can be attributed to the changing nature of the surface which contains many different types of functional groups which alter the physical properties of the surface. The calculation of the Hamaker constant is based on a flat plate model to approximate spheres. This could also contribute to the observed variations.

It should also be noted that the complexity of the surface does not lend itself well to comparison to predictions of the DLVO theory. Combined steric and electrostatic effects contribute to the overall latex stability and each stabilizing component cannot be distinguished by the experimental techniques used or any presently available theory.

CONCLUSIONS

1. The coagulum level and particle size were found to vary with AMPS concentration. As the AMPS concentration increased, a greater number of primary particles formed and were more stable. This reduces the coagulum and particle size. At high AMPS concentrations, the excess ionic comonomer homopolymerizes to form polyelectrolyte which causes flocculation most probably by bridging. This is evident through increased coagulum and particle size levels with greater concentrations of AMPS.

2. Increased concentrations of AMPS leads to greater charge on the surface. But, the higher concentrations of AMPS charged had relatively lower incorporation levels onto the particle surface.

3. Surface carboxyl groups from hydrolysis of the sulfate initiator group were found to form at acidic pH levels. When pH was controlled at approximately 7.5 as with the AMPS (Na^+) series, no carboxyl groups were found.

4. The stability of the latexes was found to increase for the AMPS (H^+) series but to follow a more random trend for the AMPS (Na^+) series.

5. The Hamaker constant was found to be of the same order of magnitude as calculated by Fowkes and reported by other workers. It was also found to vary as a function of the amount of AMPS on the particle surface and in the range of $0.5\text{-}5.0 \times 10^{-14}$ ergs for both AMPS series.

6. Ultracentrifugation studies of the latex serum showed that polyelectrolyte from the homopolymerization of AMPS could be formed.

7. The combined data from titration and stability analyses were found to correlate well with high values of surface charge giving high levels of stability. Increased surface charge led to increased

latex stability as determined by stability curve slope, c.c.c., and the Stern potential. Deviations in the expected results were attributed to the particle surface being a poor model for studying the DLVO theory. Combined steric and electrostatic effects from the unique structure of the particle surface were evident by the existing experimental deviations from the trends as predicted by the DLVO theory.

ACKNOWLEDGEMENT

This work was begun under NSF Grant No. SMI 76-03330 under an undergraduate research program. Further financial support was generously provided by members of the Emulsion Polymers Institute Industrial Liaison program.

REFERENCES

1. M.S. Juang and I.M. Krieger, J. Polym. Sci., Poly. Chem. Ed., 14, 2089 (1976).
2. B.W. Greene, D.P. Sheetz, and T.D. Fisher, J. Colloid and Interface Sci., 32, 90 (1970).
3. B.W. Greene and D.P. Sheetz, J. Colloid and Interface Sci., 32, 96 (1970).
4. B.W. Greene and F.L. Saunders, J. Colloid and Interface Sci., 33, 393 (1970).
5. Clean Surfaces: Their Preparation and Characterization for Interfacial Studies (G. Goldfinger, ed.), 15, Marcel Dekker, New York (1970).
6. N. Fuchs, Z. Physik., 89, 736 (1934).
7. J.Th.G. Overbeek and E.J.W. Verwey, Theory of the Stability of Lyophobic Colloids, Elsevier, New York (1948).

8. H. Reerink and J. Th. G. Overbeek, Disc. Farad. Soc. 66, 490 (1970).

9. G. R. Wiese and T. W. Healy, Trans. Farad. Soc., 66, 490 (1970).

10. W. W. White, in Advances in Emulsion Polymerization and Latex Technology, Short Course Notes, Lehigh University, 1977.

11. F. M. Fowkes, Ind. Eng. Chem. 56, 40 (1964).

12. A. Watillon and A. M. Joseph-Petit, Disc. Farad. Soc. 42, 143 (1966).

13. R. H. Ottewill and J. N. Shaw, Disc. Farad. Soc. 42, 154 (1966).

ALKALI-SWELLABLE LATEX: SYNTHESIS AND CHARACTERIZATION

Ronald Flaska, Syed M. Ahmed, Mohamed S. El-Aasser,
Gary W. Poehlein, and John W. Vanderhoff

Departments of Chemical Engineering & Chemistry and
Emulsion Polymers Institute
Lehigh University
Bethlehem, Pennsylvania

ABSTRACT

A series of carboxylated ethyl acrylate latexes were prepared using acrylic acid in the range of 2.1 to 12.8 mole percent under polymerization conditions of monomer "flooding". The particle size was found to increase with increasing the acrylic acid charged within the above range. Further increase in the acrylic acid resulted in complete flocculation of the latex.

The swellability of "cleaned" ion-exchanged latexes were determined by measurements of the dissymmetry ratios upon neutralization with alkali. Latexes prepared with low acrylic acid content up to 8 mole % show a net swelling of the particles upon neutralization. Where as latexes prepared with a higher content of acrylic acid exhibited a succession of swelling and disintegration cycles upon neutralization. In all cases the response of the latex particles to neutralization was found to be a function of time. From the time the pH was adjusted, 20-40 minutes were found to be required, in most cases, to reach an equilibrium diameter.

INTRODUCTION

The synthesis of alkali-swellable latexes is relatively simple, but detailed results characterizing them with respect to acid content are sparsely available. Alkali-swellable latexes are copolymers or terpolymers prepared by inclusion in the polymerization recipe of carboxyl-containing monomers such as acrylic, methacrylic, itaconic and maleic acids. The swelling of these latexes takes place upon neutralization of the carboxyl groups with alkali, which become ionized and swell because of absorption of water. This swelling usually results in a change in the physical appearance of the latex, e.g., change from milky to transparent, or a change in the rheological properties, e.g., increase in viscosity. These changes in the properties of carboxylated latexes are desirable in many applications, for example in printing inks the latexes which become transparent upon neutralization are an important class of binders. On the other hand carboxylated latexes which exhibit a change in the rheological properties upon neutralization find important application in paints, unwoven textiles and carpet backing.

The swelling behaviors of carboxylated latexes are vastly different as demonstrated in Table I, which lists various swelling properties, expressed in terms of the dissymmetry ratios, as a function of pH for a variety of industrially-produced carboxylated-latexes. Latexes 1 and 3 increase in particle size at higher pH values and the dispersions remain cloudy. Latexes 5, 6, and 7, however, decrease in particle size and become clear. Some latexes such as latex 4, decrease in particle size but do not become totally transparent. Disintegration of latex 2 occurs at a pH of 10. Latex 8 shows a slight decrease in particle size at a pH of 9, but returns to its initial particle size at a pH of 10. The type of swelling result that will be obtained for a particular latex

TABLE I

NEUTRALIZATION OF COMMERCIAL CARBOXYL-CONTAINING LATEXES

Latex	Particle Size, nm	Initial pH	Dissymmetry			Appearance
			Initial pH	pH 9	pH 10	
1	101	4.7	1.44	1.44	2.22	opaque
2	128	5.0	1.80	1.79	0.4-0.6	clear
3	84	4.8	1.27	1.29	2.15	opaque
4	160	4.3	2.58	2.18	1.46	translucent
5	94	4.8	1.36	----	1.24	clear
6	87	4.6	1.29	----	1.25	clear
7	83	4.4	1.26	----	1.17	clear
8	146	4.7	2.18	2.07	2.17	clear

is difficult to predict without systematic studies on well characterized latexes.

The neutralization with alkali and absorption of water in a carboxylated latex with high solids results in a remarkable increase in viscosity; a phenomenon which was described by Wesslau (1) as due to "inner thickening" of the dispersed latex particles. It is because of this "inner thickening" effect that the carboxylated latexes have found uses as thickening agents in a wide variety of applications.

A number of investigations (1-9) have been carried out on carboxylated latexes in order to understand the factors which control their swelling behavior. Verbrugge (9) summarized, qualitatively, the important variables which control the swellability of the carboxylated latexes as being the acid content, the hydrophilicity of the comonomers, the glass transition temperature and the molecular weight of the polymer. In order to quantify the effect of the above factors a systematic study is required to be carried out using "cleaned", well characterized carboxlated latexes. This paper reports on an initial attempt to follow the swelling behavior of "cleaned" well characterized carboxylated-latexes upon neutralization as a function of the carboxyl content of the latexes.

EXPERIMENTAL

I. Materials

Ethyl acrylate (Eastman Kodak Company, Rochester, N. Y.), was washed three times with a 10 weight percent solution of sodium hydroxide to remove the hydroquinone used as an inhibitor. The ethyl acrylate was then washed with deionized water until neutral, and dried over anhydrous sodium sulfate (reagent grade). The acrylic acid (Eastman Kodak Company), was saturated with reagent grade sodium chloride to separate the aqueous layer, dried over reagent grade calcium chloride and distilled invacuo to remove the p-methoxyphenol used as an inhibitor. The 1-dodecane thiol (Eastman Kodak Company), was purified by vacuum distillation. Doubly distilled deionized and deoxygenated water was used in all polymerizations. The potassium persulfate and sodium lauryl sulfate were reagent grade and used as received.

Caustic solutions used for washing and swelling experiments were prepared from solid sodium hydroxide (Fisher Scientific Company). Sodium hydroxide solution for conductometric titrations was obtained by dilution of 50 weight percent solutions. The 0.01875 N sodium hydroxide used for conductometric titrations was standardized against a solution of potassium hydrogen phthalate. Dowex 1-X4 (OH^--form) and Dowex 50W-X4 (H^+-form) ion exchange resins (The Dow Chemical Company, Midland, Michigan), were rigorously purified by the procedure recommended by Vanderhoff et al (10).

II. Preparation of Latexes

The latexes used in this study were prepared by emulsion polymerization using the recipes which are given in Table II. Sodium lauryl sulfate was intentionally used in order to keep the particle size of the latexes small enough within a range suitable for light scattering study. Emulsion polymerizations were carried out in a four-necked 500 ml flask kept in a constant temperature bath at 70°C. Water, potassium persulfate, and sodium lauryl sulfate were charged into the flask and remained under constant agitation using a propeller rotating at 80 rpm. The system was purged with Airco Grade-5 nitrogen, with the exit gas flowing through a reflux condenser. The monomers and mercaptan were mixed and added to the contents of the flask via a 100 ml burette at a rate of 0.9 ml/min, the addition time was 1.5 hours. Under these conditions the reaction flask was flooded with monomer.

TABLE II

RECIPES FOR ETHYL ACRYLATE-ACRYLIC ACID COPOLYMERS

Latex	1A	2A	3A	4A	5A	6A	7A*	8A*
Mole Percent Acid (Based on Monomer)	2.1	4.2	6.3	8.0	10.6	12.8	15.0	17.2
Ethyl Acrylate (gm)	75.40	74.22	73.01	72.06	70.52	69.23	67.92	66.58
Acrylic Acid (gm)	1.16	2.34	3.55	4.50	6.04	7.33	8.64	9.98
(meq)	16.11	32.5	49.3	62.5	83.9	101.8		

1 dodecane-thiol	0.16 gm.
water	119.98 gm.
potassium persulfate	0.22 gm.
sodium lauryl sulfate	3.08 gm.
monomer	76.56 gm.

*latexes coagulated

The entire polymerization, starting from the point where monomer addition was begun, proceeded for 18 hours. Usually the polymerizations were carried out to 95+% conversion; the remaining monomers were removed from the latex by steam distillation under vacuum with a Fisher Flash evaporator.

III. Purification and Characterization of Latexes

The latex was diluted to 5% solids with deionized water and ion-exchanged with mixed-bead resins of Dowex 1 and Dowex 50. The ion-exchange was repeated six times, with 4-6 hours contact time in each cycle. The ion-exchanged latexes, with their acidic groups in the H^+-form, were characterized with respect to the carboxylic acid content by conductometric titration against alkali. The conductometric titrations were performed using the apparatus constructed by Wu (11). An accurately determined weight of ion-exchanged latex (approximately 0.30 gm), was diluted to 140 ml in a 150 ml beaker and placed under constant agitation using a magnetic bar. A 0.01875 N sodium hydroxide solution was introduced into the latex via a burette at a flow rate of 0.8 ml/min. The conductivity of the latex was determined by two platinum-leaf electrodes supported on a pair of holders which were placed in a beaker. A conductivity analyzer connected to the electrodes produced a voltage in direct proportion to the solution's conductivity and the output was recorded.

IV. Swelling of Latexes

A liter of 1 x 10^{-3} weight percent solids were prepared from each ion-exchanged latex by dilution using doubly-distilled deionized water. A Brice Phoenix Universal Photometer was used in conjunction with a photomultiplier to obtain swelling measurements. A D-102 type cell was employed to hold the latex to be studied and was filled with 100 ml of the diluted latex. The intensity of scattered light at 45° and 135° to the exiting direct beam was measured and the dissymmetry ratios were calculated. The light wavelength was 546 nm. Particle sizes were obtained from the dissymmetry ratios using tables based on the work by Beattie and Booth (12).

Dissymmetry ratios were measured before the latexes were swelled at a pH of about 5.5. The pH was then adjusted by adding the caustic solution and using a Fisher pH-meter Model 230, to follow the pH. Timing was begun after the addition of sodium hydroxide, and dissymmetry ratios were measured every 5 minutes until a constant reading was obtained. The change in the dissymmetry ratios with time for each latex system was studied at pH values of 9, 10, and 11.

RESULTS AND DISCUSSION

I. Surface Characterization of "Cleaned" Latexes

Figure 1 shows the conductometric titration curves for all six ion-exchanged "cleaned" latexes. All curves are displaced on the conductivity axis, for clarity. The first end point at the bottom of the descending leg in all curves, is used to calculate the quantity of the strong acid groups. The concentrations of strong acid groups were found to be in the range of 5.7-to-9.1 x 10^{-6} gm/gm polymer. These strong acid groups are the sulfate end groups from the persulfate initiator.

The ascending legs of the titration curves for latexes 1A and 2A show two slopes; the first is due to the titration of the carboxyl groups and the second slope is due to excess sodium hydroxide. The end point between the two lines is used to calculate the amount of the carboxyl groups titrated in the latex under consideration. The results are given in

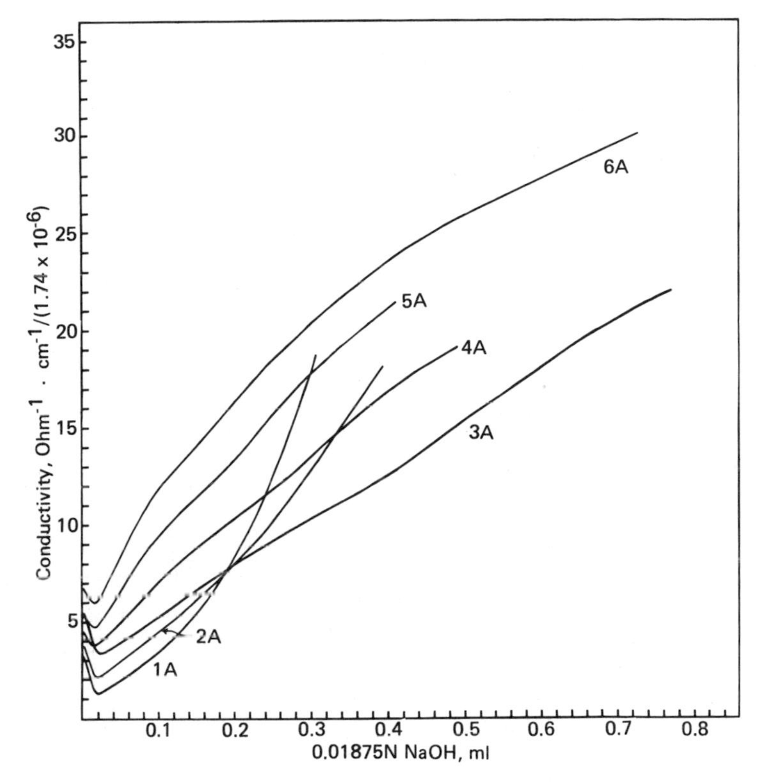

Figure 1. Conductometric titration curves of ion-exchanged carboxylated latexes.

Table III. The concentrations of carboxyl groups titrated were 95 x 10^{-6} (Latex 1A) and 120 x 10^{-6} (Latex 2A) μ eq/gm polymer. These concentrations represent 45.1% and 28.3% of the acrylic acid charged in the polymerization recipes of the two latexes.

The ascending legs in the titration curves of latexes 3A, 4A, and 5A are comprised of several parts which alternate between increasing and decreasing slopes. The titration curve of latex 6A is different from the other curves in that the first part of the ascending leg is followed by other parts with continuously decreasing slopes. In all four latexes, however, the slopes of the final part of the ascending leg

TABLE III

CHARACTERIZATION RESULTS OF CARBOXYLATED LATEXES

	Charge Density, μeq/gm polymer					Acrylic Acid charged	% of Acrylic Acid in
	Strong	Weak Acids					Polymer Relative
Latex	Acid	1st	2nd	3rd	Total	meq/gm polymer	to Charge
1A	9.1	95	---	---	95	0.210	45.1
2A	8.2	120	---	---	120	0.424	28.3
3A	9.6	90	110	110	(310) Incomplete	0.644	
4A	5.7	50	56	64	(170) Incomplete	0.816	
5A	8.3	42	51	58	(151) Incomplete	1.096	
6A	8.9	44	170	---	(214) Incomplete	1.330	

is not near the expected slope of the line representing excess sodium hydroxide. This is an indication of incomplete titration of all carboxyl groups.

Several reasons were considered as possible explanations of the shapes of the conductometric titration curves in Figure 1. The first consideration was given to the possibility that a continuous hydrolysis of ethyl acrylate is taking place during the course of the titration of the ion-exchanged latex with sodium hydroxide causing the generation of new carboxyl groups. Hydrolysis of acrylate esters is known to take place with a rate which is a function of temperature, degree of tacticity of the polymer, and percent of neutralization of the carboxylic groups within the copolymer (13). This explanation can be discredited on two grounds. First, the duration of the conductometric titration was short (5-10 minutes), and was conducted at 25°C which makes the hydrolysis of ethyl acrylate negligible. Second, the experimental results shown in Table III indicate that the amounts of the carboxyl groups titrated in the "cleaned" latexes 1A and 2A are much less than the amounts of the acrylic acids originally charged. It should be mentioned that the conductometric titrations which were carried out at a later date on all six latexes after storage for 5 months in the ion-exchanged form, gave results which were completely different from the titration results of the freshly prepared and ion-exchanged latexes of Table III. Conductometric titrations were carried out on all latexes using a very small weight of polymer (about 0.03 gm). The results given in Table IV indicate that the carboxyl contents in the aged latexes (1A and 2A) were much higher than the carboxyl contents in the freshly prepared and ion-exchanged latexes of Table III. The carboxyl contents titrated in the five latexes, 2A-6A, were much higher than the acrylic acid charged in the recipe; even though part of the water-solubel carboxylated polymer was

TABLE IV

CHARACTERIZATION OF ION-EXCHANGED CARBOXYLATED LATEXES STORED FOR 5 MONTHS IN THE H^+-FORM

Latex	Wt. of Polymer titrated (gm)	Total Carboxyl content meq/gm polymer	Acrylic Acid charged meq/gm polymer	% of Acrylic Acid in Polymer Relative to Charge
1A	0.0488	0.167	0.210	79.5
2A	0.0472	0.430	0.424	101.4
3A	0.0542	0.778	0.644	120.8
4A	0.0208	1.191	0.816	146.0
5A	0.0241	1.589	1.096	145.0
6A	0.0271	1.635	1.330	122.9

probably removed during the ion exchange cycles. This is a clear indication that the hydrolysis of the ethyl acrylate is catalyzed when the latex is stored in the acid form. This point requires further investigation. However, all latexes which were used in the initial characterization and in all the swelling studies were freshly prepared, ion-exchanged latexes. Another possible reason for the shape of the titration curves in Figure 1, is that during the 5-10 minute duration of the titration, equilibrium is not attained. This could result because some of the carboxyl groups are buried within the core of the particles. This point requires further experiments with conductometric titrations at slower rates of addition of the sodium hydroxide to the ion-exchange latexes.

The third possible explanation for the shape of the titration curves is the idea suggested earlier by Muroi (4), that upon the addition of alkali to a carboxylated latex, the outer layer is swelled due to neutralization of the carboxyl groups. This is then followed by disintegration or "peeling", as of an onion skin, of this outer layer. One can imagine a succession of swelling-peeling cycles to take place during the neutralization of highly carboxylated latexes. Each time peeling of the outer layer takes place, a new layer with carboxyl groups will be exposed which in turn will be neutralized, possibly giving a conductometric titration curve with different slope. The swelling results reported in the next section indeed suggest that the last reason is a valid one which could be used in explaining both the shape of the conductometric titration curves as well as the swelling results.

II. Particle Size and Swelling Behavior

The particle sizes of the "cleaned" unswelled latexes determined from dissymmetry ratios, are plotted in Figure 2 as a function of the mole percent of the acrylic acid. The particle size increases with increasing acrylic acid content; with a remarkable sudden increase in size above 6.3 mole percent of the acid. The steep slope of this curve above 6.3 mole percent acrylic acid suggests that aggregation of the particles play an important role in controlling the particle size of these latexes. Aggregation could be caused by water-soluble polymers which can be formed during the polymerization of water-soluble monomers such as acrylic acid, especially under the monomer "flooding" conditions which prevailed during the preparation of the present latexes. Latexes 7A and 8A, prepared with 15 and 17.2 mole percent of acrylic acid respectively, aggregated completely at the end of the polymerization; most probably due to excessive flocculation by water-soluble polymers.

The swelling results of the ion-exchanged latexes are given in Figures 3-8, which show the change in particle diameter with time as a function of pH. The swelling results

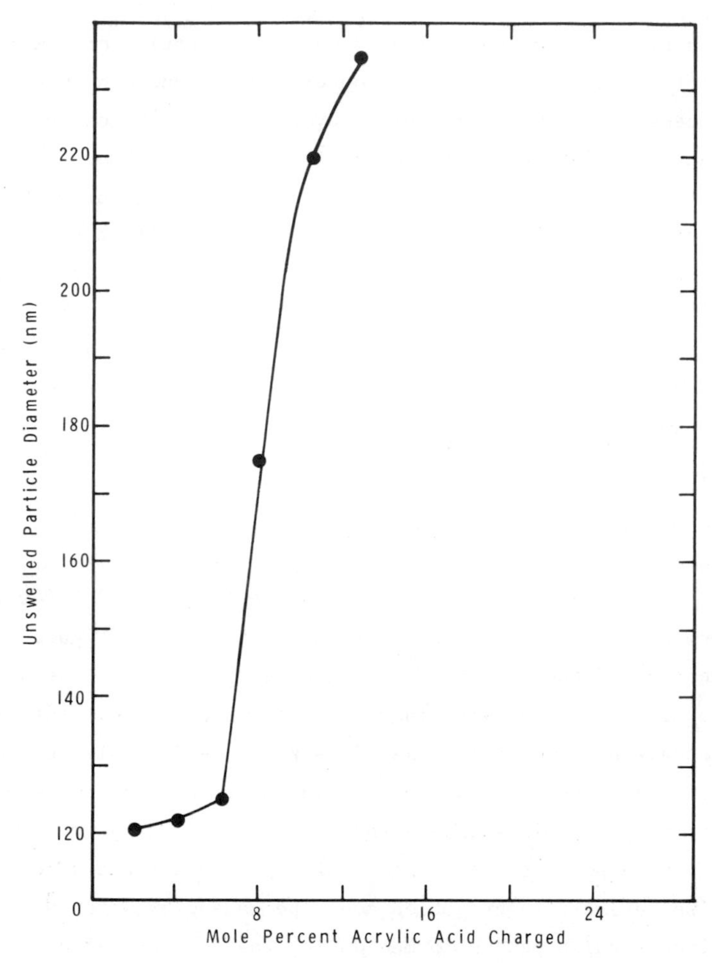

Figure 2. Latex particle diameter as a function of mole percent acrylic acid charged in the polymerization recipe.

are also summarized in Figure 9 and Table V, where $\underline{D}_i$ is the initial latex particle diameter determined from dissymmetry ratios measured at pH about 5.5 before adding the alkali, $\underline{D}_f$ is the final latex particle diameter taken at the point where the dissymmetry ratio reaches a constant value usually 40-60 minutes after the adjustment of the pH; ΔD is = (D_f-D_i) the percent swelling with respect to the initial diameter, is

TABLE V

SWELLING RESULTS

Latex	pH	D_i(nm)	D_f(nm)	ΔD	% Swelling
1A	10.0	121	120	-1	-0.8%
	11.0	121	126	5	4.1%
	12.0	121	120	-1	-0.8%
2A	9.0	122	122	0	0.0%
	10.0	122	141	19	15.6%
	11.0	122	142	20	16.4%
3A	9.0	125	123	-2	-1.6%
	10.0	125	150	25	20.0%
	11.0	125	152	27	21.6%
4A	9.0	175	175	0	0.0%
	10.0	175	212	37	21.1%
	11.0	175	221	46	26.3%
5A	9.0	220	211	-9	-4.1%
	10.0	220	186	-34	-15.5%
	11.0	220	203	-17	-7.7%
6A	9.0	235	184	-51	-21.7%
	10.0	235	176	-59	-25.1%
	11.0	235	191	-44	-18.7%

calculated from ($\Delta D/D_i$ x 100) and is given in the last column with a negative sign to indicate shrinkage or a decrease in diameter and a positive sign to indicate swelling or an increase in diameter.

Figure 3 shows the swelling results of latex 1A prepared with the lowest acrylic acid content (2.1 mole percent), very slight swelling (4.1%) is noticed at pH 11 and a negligible shrinkage (less than 1%) in size takes place at pH 10 and 12.

Figures 4, 5, and 6 show the swelling behavior of ion-exchanged latexes, 2A, 3A, and 4A prepared with 4.2, 6.3, and 8 mole percent acrylic acid respectively. In these three latexes the initial particle size remained almost constant after adjusting the pH to 9, except latex 3A which exhibited a

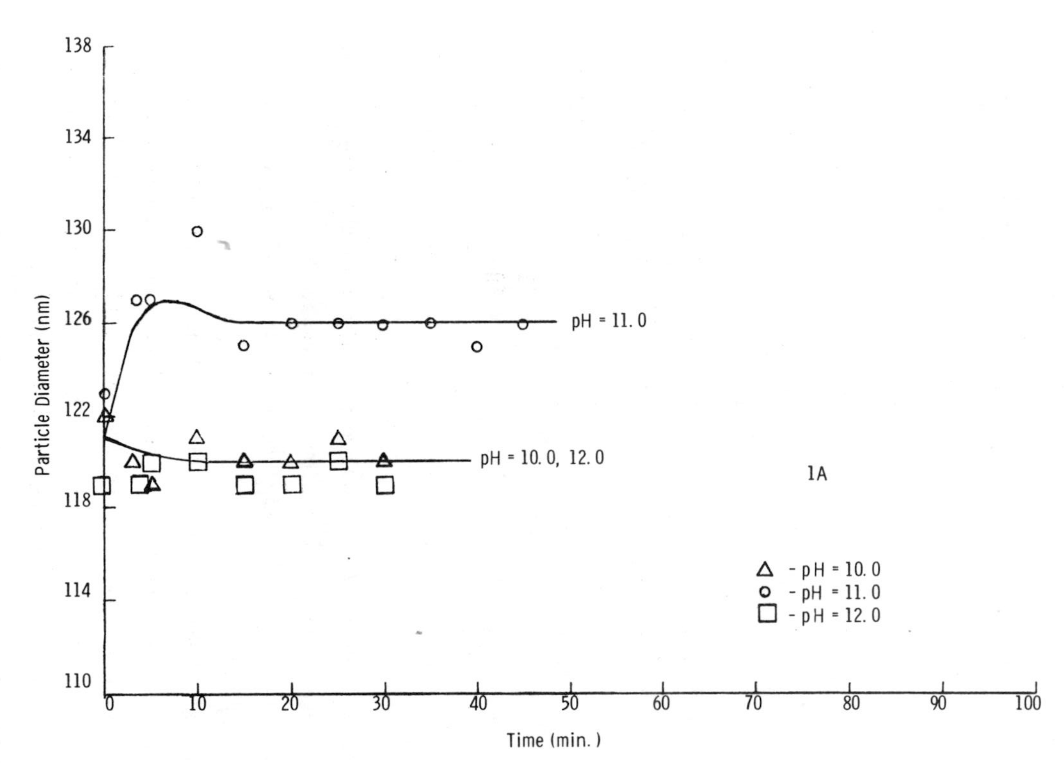

Figure 3. pH effect on the variation of particle diameter with time for latex 1A prepared with 2.1 mole percent acrylic acid.

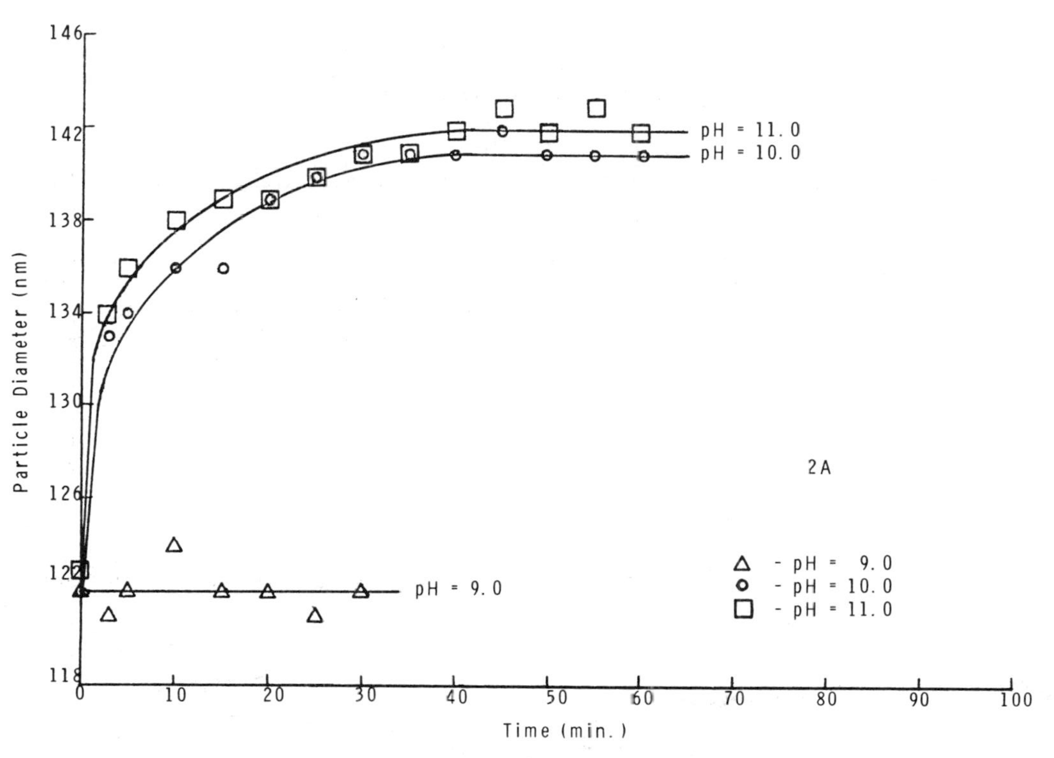

Figure 4. pH effect on the variation of particle diameter with time for latex 2A prepared with 4.2 mole percent acrylic acid.

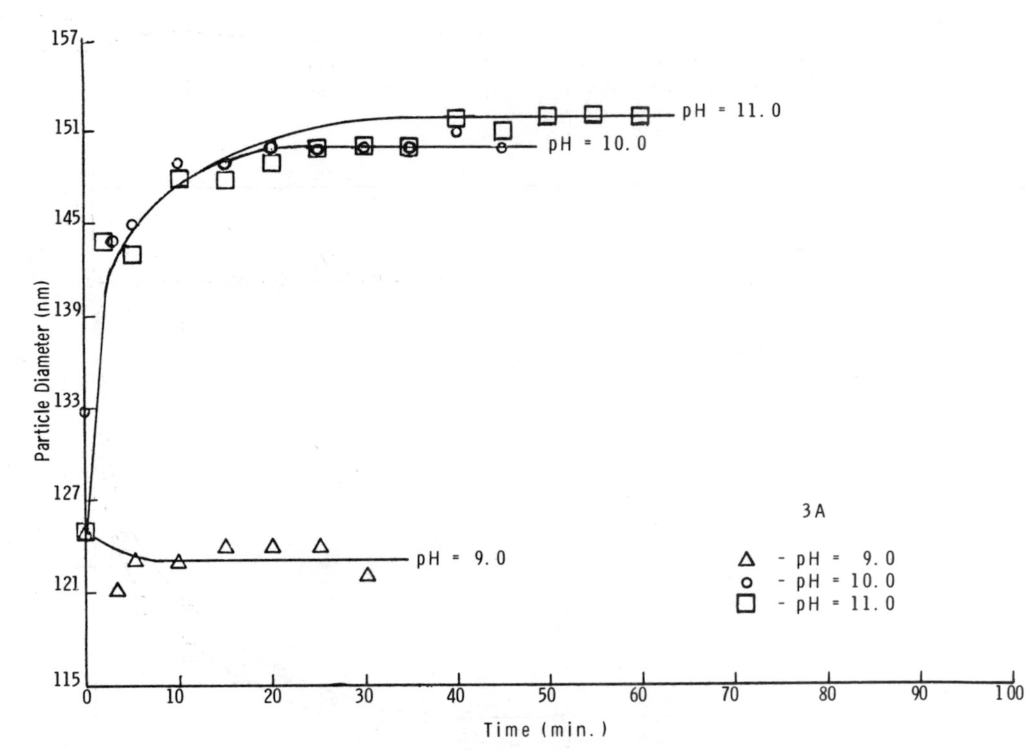

Figure 5. pH effect on the variation of particle diameter with time for latex 3A prepared with 6.3 mole percent acrylic acid.

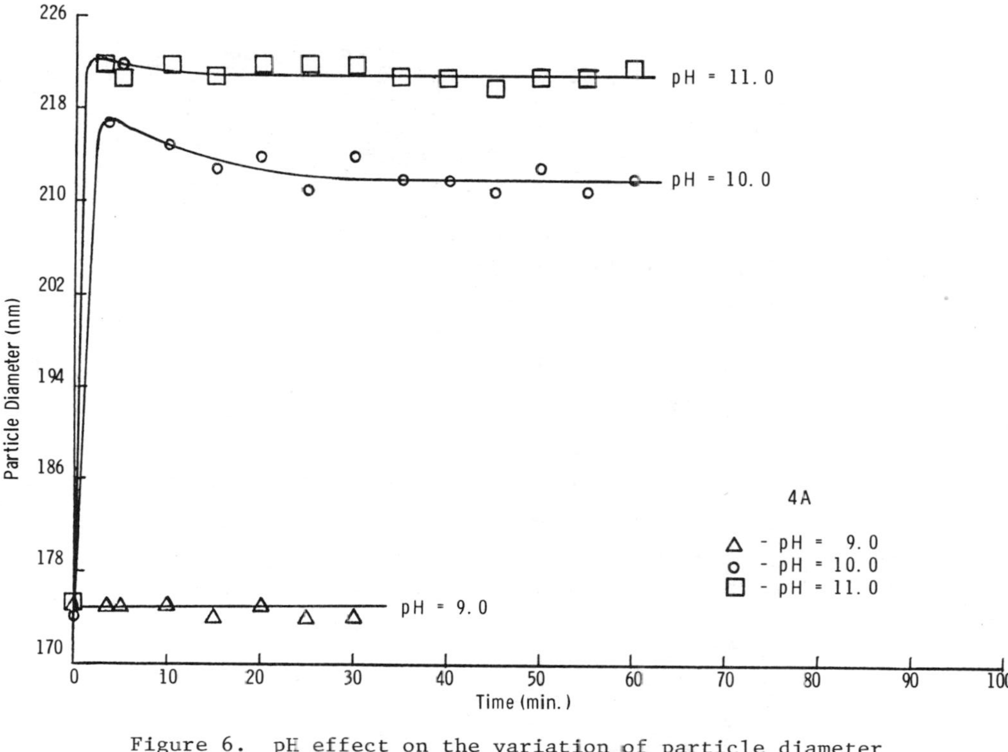

Figure 6. pH effect on the variation of particle diameter with time for latex 4A prepared with 8.0 mole percent acrylic acid.

slight shrinkage of 1.6%. The pH's of these latexes measured after 60 minutes from the time their pH was adjusted to 9, were found to be 8.4. The decrease in the pH is due to the neutralization reaction of the carboxyl groups buried within the particles, which should result in an increase of the particle size. However, the experimental result that no change in particle size was found upon neutralization to pH 9, could be explained by a balance between an increase in size of the particle due to neutralization and a shrinkage of its size which usually results from the addition of an electrolyte, such as sodium hydroxide, to ion-exchanged latex. The slight decrease in the particle size of latex 3A indicates that upon neutralization to pH 9 the shrinkage outweighed the swelling of the particles.

On the other hand, when the pH of these three latexes was adjusted to higher values of 10 and 11 by adding alkali, a different swelling behavior was observed. Figures 4-6 show that upon adjusting the pH to a given value, the particle size increases as a function of time and 20 to 40 minutes are needed before an equilibrium particle diameter is reached. The three latexes, however, exhibit various degrees of swelling as given in Table III. In all three latexes the percent swelling at pH 11 is higher than at pH 10. At a given pH, the percent swelling increases as the percent of acrylic acid charged increases. This swelling behavior is the result of the neutralization reaction of the carboxyl groups with the added alkali. The time dependency of the swelling is most probably due to the diffusion of the alkali to the core of the particles through the swollen outer layers. In support of the time required for the alkali to diffuse into the core of the latex particles, it was found that the pH measured 60 minutes after its initial adjustment decreased from 10 to 9.7 and from 11 to 10.8 in all three latexes.

Latex 4A prepared with 8 mole percent acrylic acid exhibited a slightly different swelling behavior than the other two latexes. At high pH, the particles appear to swell to a high level and then shrink slightly before becoming constant after about 40 minutes. This may indicate the beginning of a noticeable "peeling" of the swollen outer layers of the neutralized carboxylated latex particles.

Figures 7 and 8 show the swelling behavior of ion-exchanged latexes 5A and 6A prepared with 10.6 and 12.8 mole percent acrylic acid respectively. Similar swelling behaviors are seen in the two latexes where the particle size decreases with time at the three pH values, and about 40 minutes are needed to reach an equilibrium diameter.

The equilibrium particle diameter at pH 11 is always larger than that at pH 10 for the two latexes. The above results suggest that upon neutralization at pH 9 and 10 of carboxylated latexes with high acid content, the outer layers peels off the surface with the result of decreasing the particle size. This in turn results in exposing the inner carboxyl groups which neutralize upon further addition of alkali (pH = 11), with the result of swelling the particles to larger size.

CONCLUSIONS

1. The particle size of the carboxylated ethyl acrylate latexes was found to increase with the mole percent of the acrylic acid charged. Flocculation, most probably due to bridging by water-soluble polymers, seems to be the cause of the increase in particle size. Further increase in the acrylic acid content above 14 mole percent caused complete coagulation of the latex.

2. The titration of the carboxyl groups in the "cleaned" latexes accounted for 45.1% of the 2.1 mole percent acrylic acid used in the preparation of the latex, and only 28.3% when 4.2 mole percent acrylic acid was used.

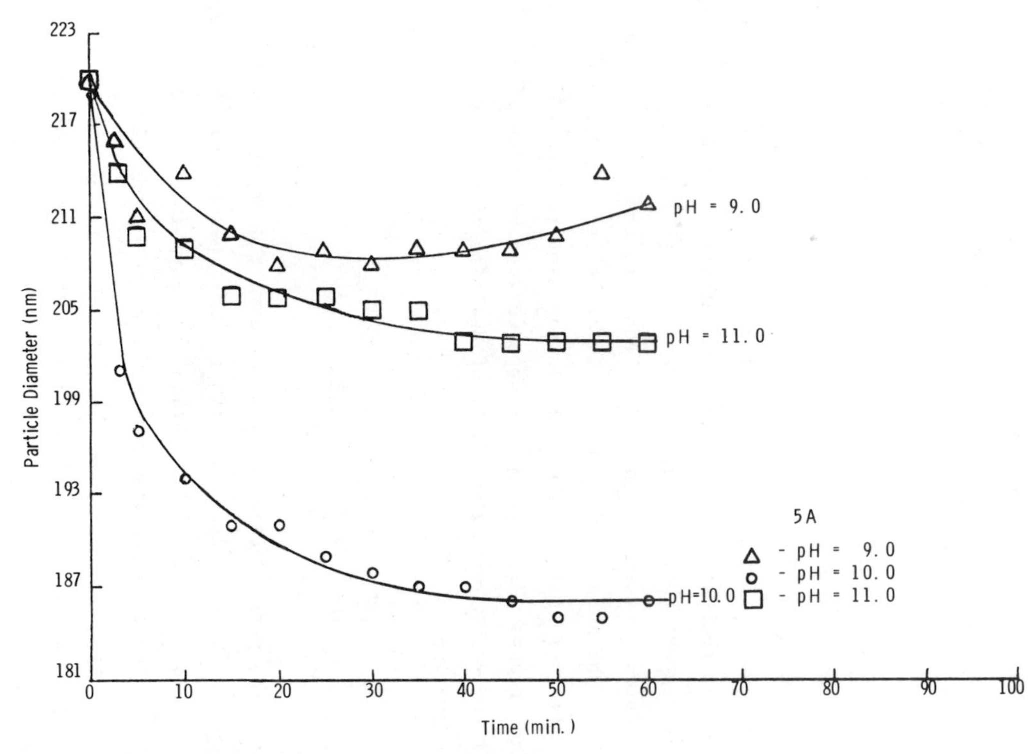

Figure 7. pH effect on the variation of particle diameter with time for latex 5A prepared with 10.6 mole percent acrylic acid.

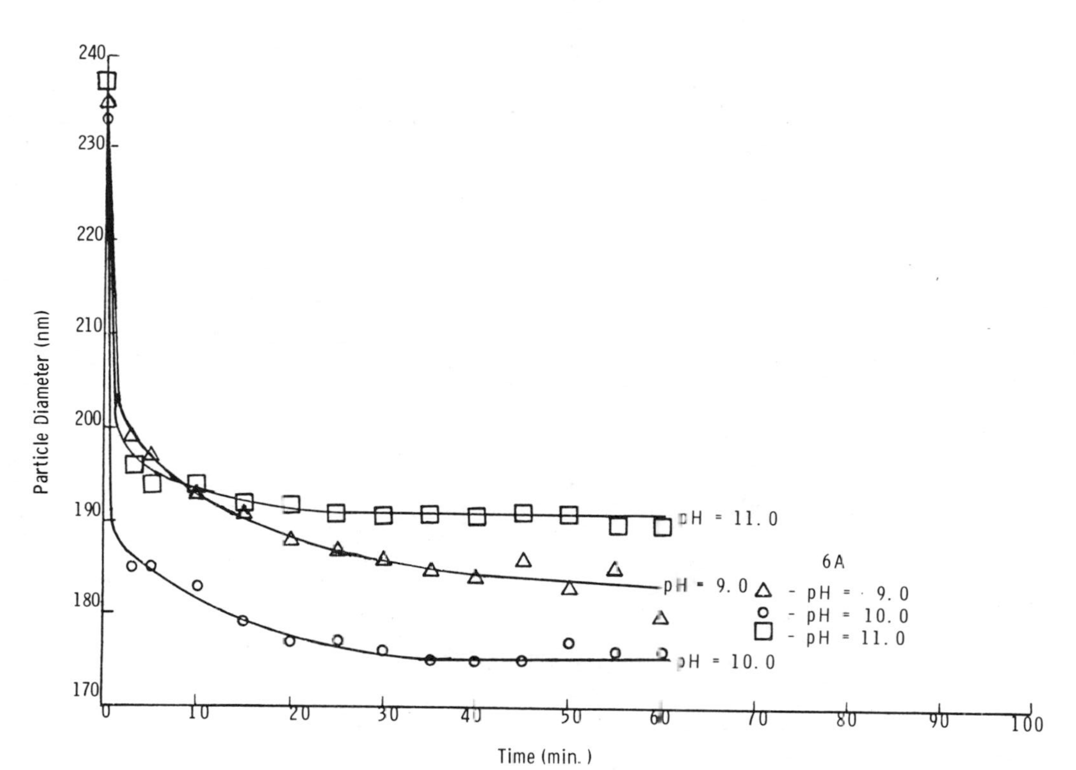

Figure 8. pH effect on the variation of particle diameter with time for latex 6A prepared with 12.8 mole percent acrylic acid.

3. The titration of the carboxyl groups in the "cleaned" latexes prepared with mole percent acrylic acids in the range of 6.3-12.8, were incomplete. The titration curves exhibit alternating lines with increasing and decreasing slopes. Hydrolysis of ethyl acrylate, unattained equilibrium during titration due to neutralization of inner carboxyl groups, and alternation between swelling and peeling-off of outer layers

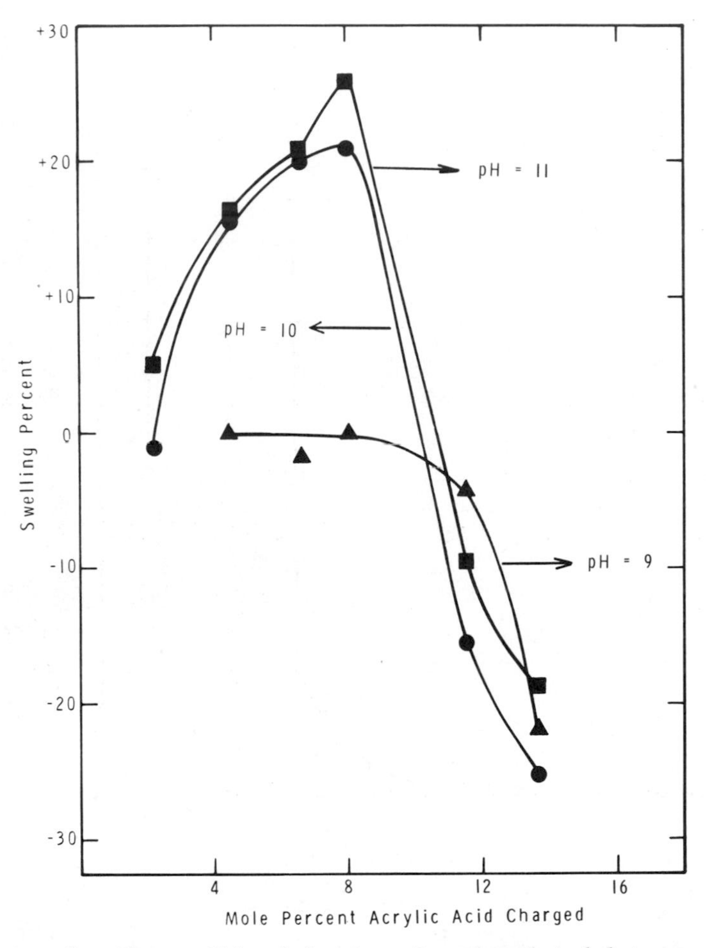

Figure 9. The swelling behavior of carboxylated latexes at various pH values as a function of mole percent acrylic acid charged in the polymerization recipe.

were given as possible causes for the shapes of the titration curves.

4. The swelling behavior of the ion-exchanged carboxylated latexes are summarized in Figure 9, which shows the change in the percent swelling as a function of the mole percent of the acrylic acid at three pH values of 9, 10 and 11. The neutralization of the latexes to pH 9 resulted in neither swelling nor shrinkage of their particles up to 8 mole percent acrylic acid. However, further increase in the acid content resulted in a definite shrinkage of the particles due to disintegration of the outer swollen layers.

The neutralization of the carboxylated latexes to pH 10 and 11 resulted in a steady swelling which increased with the increase in the acrylic acid content up to 8 mole percent. Also, in this acid range the percent swelling of the particles was larger at pH 11 than at pH 10 due to higher degree of neutralization of the carboxyl groups at the highest pH value. The increase in the acid content above 8 mole percent resulted in a drastic shrinkage of the particle size due to peeling off of the outer swollen layers. At these higher levels of acid contents the particle size at pH 11 were consistently larger than those at pH 10 due to the further swelling of the particles, after disintegration of the outer layer and upon further neutralization to pH 11.

ACKNOWLEDGEMENT

This work was begun under NSF Grant No. SMI 76-03330 under an undergraduate research program. Further financial support was generously provided by members of the Emulsion Polymers Liaison Program.

The authors are deeply indebted to G. Pauli and D. Houston for their invaluable contributions to this work.

REFERENCES

1. H.·Wesslau, Makromol. Chem., 69 220 (1963).
2. D. B. Fordyce, J. Dupre and W. Toy, Offic. Dig., 31, 284 (1959).
3. D. B. Fordyce, J. Dupre and W. Toy, Ind. Eng. Chem., 51, 115 (1959).
4. S. Muroi, J. Appl. Polymer Sci., 10, 713 (1966).
5. S. Muroi, J. Appl. Polymer Sci., 11, 1963 (1967).
6. K. Tyuzyo, H. Harada and H. Morita, Kolloid Z. Polymere, 201, 155 (1965).
7. L. F. Guziak and W. N. Maclay, J. Appl. Polymer Sci., 7, 2249 (1963).
8. C. J. Verbrugge, J. Appl. Polymer Sci., 14, 897 (1970).
9. C. J. Verbrugge, J. Appl. Polymer Sci., 14, 911 (1970).
10. J. W. Vanderhoff, H. J. van den Hul, R. T. M. Tausk, and J. Th. G. Overbeek, "Clean Surfaces: Their Preparation and Characterization for Interfacial Studies", G. Goldfinger, editor, Marcel Dekker, New York, 1970, p. 15.
11. W.-C. Wu, Ph.D. Thesis, Lehigh University, 1977.
12. W. H. Beattie and C. Booth, J. Phys. Chem., 64, 696 (1960).
13. "Chemical Reactions of Polymers", E. M. Fettes, editor, High Polymers, Vol. 24, Interscience, New York, 1964.

PARTICLE SIZE DISTRIBUTION OF COLLOIDAL LATEXES BY HDC

Cesar A. Silebi
Anthony J. McHugh

Department of Chemical Engineering
and
Emulsion Polymers Institute
Lehigh University
Bethlehem, Pennsylvania

1. INTRODUCTION AND BACKGROUND

The determination of particle size distributions (PSD) in the colloidal size range is an extremely important facet of polymer latex technology. The optical, rheological, and electrical properties (to name a few) as well as stability of a latex are profoundly affected by the PSD. In our labs we have been pursuing the development of chromatographic methods for the determination of colloidal latex size distributions. It is our belief that such methods will ultimately be applied as routine analytical tools for rapid and reliable determination of a latex PSD in the several hundred to several thousand Angstroms diameter range. Two variants on the basic method, HDC (1), and LEC (2,3), have been reported. This paper will be addressed to the work in our labs on HDC. Our results will be presented in essentially two parts, the first dealing with fundamental analyses of the separation mechanism and the second with a discussion of methods developed for the direct determination of the PSD from the HDC chromatogram of a polydisperse system.

The experimental details and instrument have been described elsewhere (4) as well as in the studies by Small and coworkers (1,5).

The relevant experimental parameters involved in the separation process have been shown to be: ionic strength of the eluant, surfactant species and concentration, eluant flow rate, and packing diameter, as well as the latex particle size (1,6). Particle separation in the columns can be quantified in terms of the separation factor R_F (1) or equivalently the elution volume difference ΔV between particle and eluant marker turbidity peaks (6). The two are related by

$$\Delta V = V_m(1 - 1/R_F)$$

with V_m the elution volume associated with the marker peak. Figures 1 to 3 illustrate the dependence of the particle size - ΔV

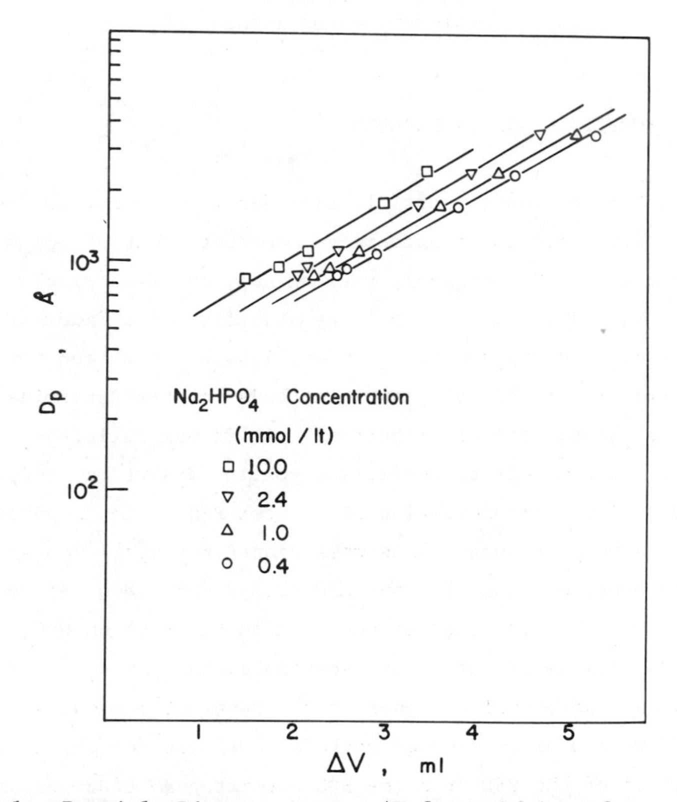

Figure 1. Particle Diameter versus ΔV for a Series of Monodisperse Polystyrene Latexes at Several Electrolyte Concentrations. (Surfactant Concentration was held constant at 2.67×10^{-3}M).

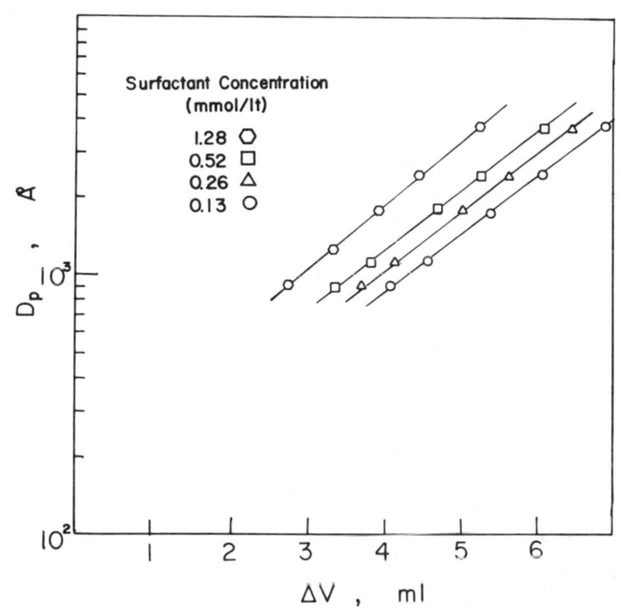

Figure 2. Particle Diameter versus ΔV for a series of Monodisperse Polystyrene Latexes at Several Emulsifier Concentrations. (Added salt (NaCℓ) held constant at 2×10^{-4}M)

relationship on two of the above mentioned parameters and demonstrate several important features. First, column resolution goes as the inverse slope of the log D_p - ΔV calibration curve, thus the resolution is independent of ionic strength, both of the surfactant and the added salt (6). Second, as Figure 2 illustrates, the surfactant, below its critical micelle concentration, acts in the separation merely as another ionic species. A similar effect is seen with AMA (Aerosol MA) surfactant as well as with PVC latexes (7) and data for the several systems can be superposed by plotting at the ionic strength which would give the same double layer thickness (8). As one increases ionic strength the log curves eventually show a non-linear behavior and reversal of the separation process with small particles eluting ahead of larger ones (6). Figure 3 shows that at a given ionic strength changes in the flow rate do not affect the calibration curve and consequently column resolution remains constant over the range studied.

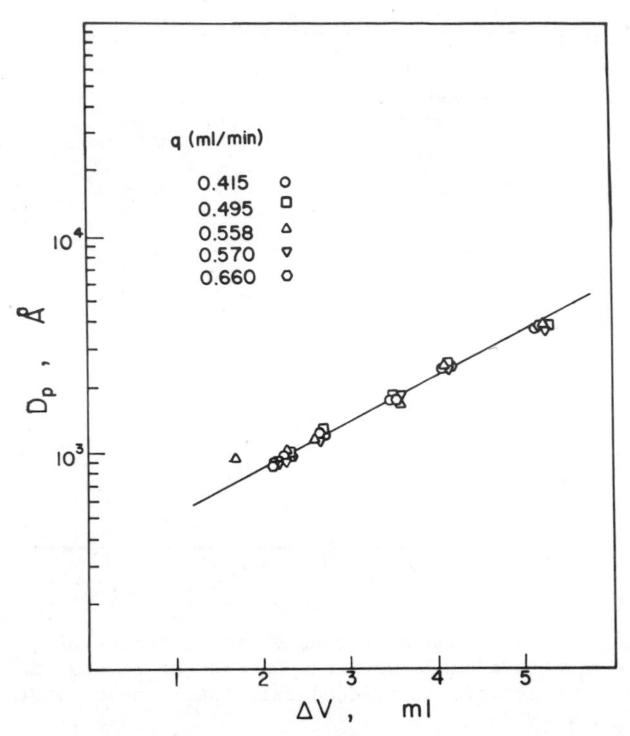

Figure 3. Particle Diameter versus ΔV for a Series of Monodisperse Polystyrene Latexes for Several Different Eluant Flow Rates.

II. ANALYSIS OF THE SEPARATION MECHANISM

Our analysis of the separation mechanism is based on the capillary tube model for the bed hydrodynamic behavior which assumes the interstitial regions between packing can be treated as an equivalent array of parallel capillaries of equal radius (4,6). The evaluation of the separation factor R_F resolves to calculating the average velocity ratio between the particle and marker species in the capillary (10). The equivalent capillary radius R_o is computed from the bed packing D_{pk} and void volume ε_v (6) and the average particle velocity in the capillary is given by (9)

$$[v_p] = \frac{\int_o^{R_o - R_p} v_p(r) e^{-\phi/kT} \, rdr}{\int_o^{R_o - R_p} e^{-\phi/kT} \, rdr} \tag{2}$$

The exponential function contains in addition to the Boltzmann constant, k, and temperature, T, the total energy of interaction, ϕ, between the particle and the capillary wall. This term represents in effect, the probability that a particle will occupy a given flow stream line at the radial position r (10). The expression for the marker velocity results by setting $R_p = o$ in equation 2 (10).

$$v_p(r) = v_o\left(1 - \frac{r^2}{R_o^2}\right) - \gamma \, v_o \left(\frac{R_p}{R_o}\right)^2 \tag{3}$$

The wall effect parameter, γ, is fit to the data of Goldsmith and Mason (11) in the core region and calculated from an expression by Goldman et al. (12) in the wall region. Thus

$$\gamma = \begin{cases} \frac{2}{3}\left(1 + 3\frac{r}{R_o}\right) & \text{core region} \quad (3a) \\ \frac{5}{8} \dfrac{\frac{R_p}{R_o}}{\left(1 - \frac{r}{R_o}\right)^2} & \text{wall region} \quad (3b) \end{cases}$$

The total energy of interaction is given as the sum of the double layer, Born, and Steric repulsion terms and the Van der Waals attraction energy. Thus

$$\phi = \phi_{st} + \phi_{D_L} + \phi_B + \phi_{vw} \tag{4}$$

The latter three terms are calculated by assuming a sphere-plane interaction and can be evaluated over the full range of surface potentials and κR_p values (κ^{-1} the double layer thickness) by standard means (10). In the case where long chain surface active agents are adsorbed on the surface of the particle and the packing a steric repulsion between the colloidal particle and the wall will arise with a contribution to the potential of interaction given by (8)

$$\phi_{st} = \frac{8\pi k\ T\ C^2}{3\rho^2\ V_2} (\psi_1 - \chi_1)\ (\delta - \frac{H_o}{2})\ (3R_p + \delta + H_o) \quad (5)$$

for $H_o \leq 2\delta$ and zero otherwise. The terms in ψ_1 and χ_1 account for the entropy of mixing and interaction energy between the stabilizer and solvent. The concentration of surfactant in the adsorbed layer, c, is obtained from the adsorption curve, ρ is the density of the adsorbed specie, V_2 is the molecular volume of the solvent. The minimum distance between the hard particle and the wall surfaces is given by H_o and δ is the thickness of the adsorbed layer. Equation (5) was derived following the development of Ottewill et al. (13) for the steric interaction between two spherical particles.

Values for the material parameters used in our calculations were chosen from ranges given in the literature as typical for polymer latex particles. Figures 4 and 5 illustrate the important role of ionic strength on the separation factor-particle diameter calculation in our model. Figure 4 shows that at the relatively high ionic strength of 0.1M (equivalent (NaCℓ) the separation behavior is highly sensitive to the value chosen for the Hamaker constant and further indicates that under these conditions separation will be controlled by the particle chemistry (i.e. value of the Hamaker constant for the system). In addition to showing that for a given latex system, particle separation may begin to reverse itself at these high ionic strengths, these calculations also indicate an alternate and important interpretation. Namely, at high ionic strengths, particles of the same diameter but differing in chemistry can be separated by HDC. This very behavior has

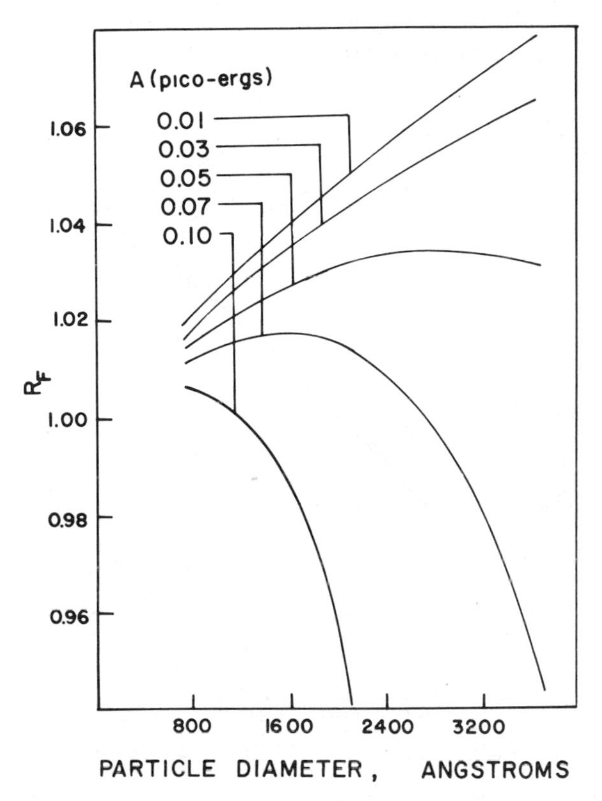

Figure 4. R_F versus D_p Calculated From the Capillary Model at 0.1M total Ionic concentration for Several Values of the Hamaker Constant.

in fact been observed in a reported series of experiments by Small (14) wherein mixtures of polystyrene and polymethylmethacrylate of nearly the same diameter were separated at the ionic strength of 0.4 M using an HDC column setup.

Results of the calculations such as shown in Figure 5 illustrate that at low enough ionic strength the separation process is indeed dominated by hydrodynamic considerations along with the dominant electrostatic repulsion. This, of course, indicates the use of low ionic strength for calibration and operation to enable use of data on a well characterized latex for universal calibration purposes. Very similar results to those indicated in

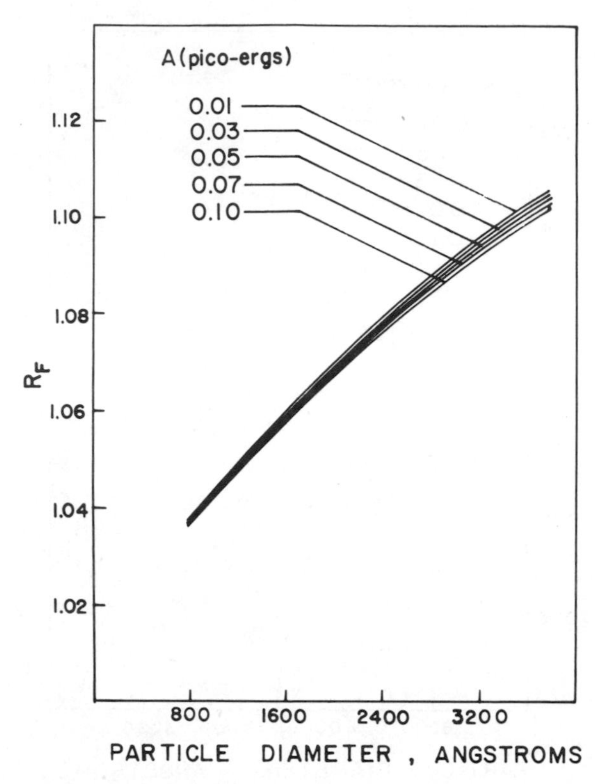

Figure 5. R_F versus D_p Calculated From the Capillary Model at 0.001M total Ionic concentration.

Figures 4 and 5 result with surface potential as a variable (10). Figure 6 illustrates the good comparison between R_F - D_p calculations for various packing sizes and the data of Small (1). Figure 7 shows a comparison between the column data of Small (1) and our calculations for the reported total ionic strengths as shown. Values for the material parameters used in the calculations were taken solely from the literature for the conditions considered (10). In addition, the capillary radius was calculated from data for Small's columns. The comparison with Small's data is thus the result of a zero free parameter calculation and shows rather strikingly the close agreement which can be obtained with our capillary model. A much closer fit can be obtained in the

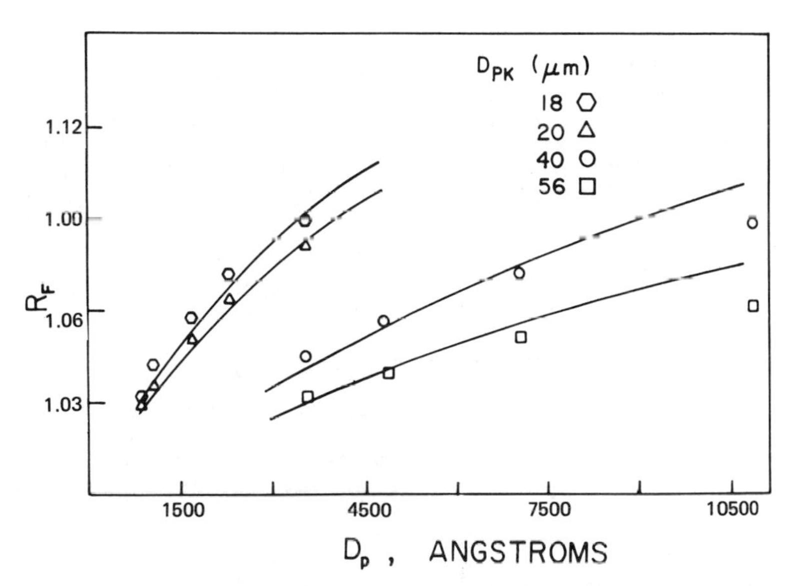

Figure 6. Comparison between calculated (solid line) and measured (1) values for various packing diameters.

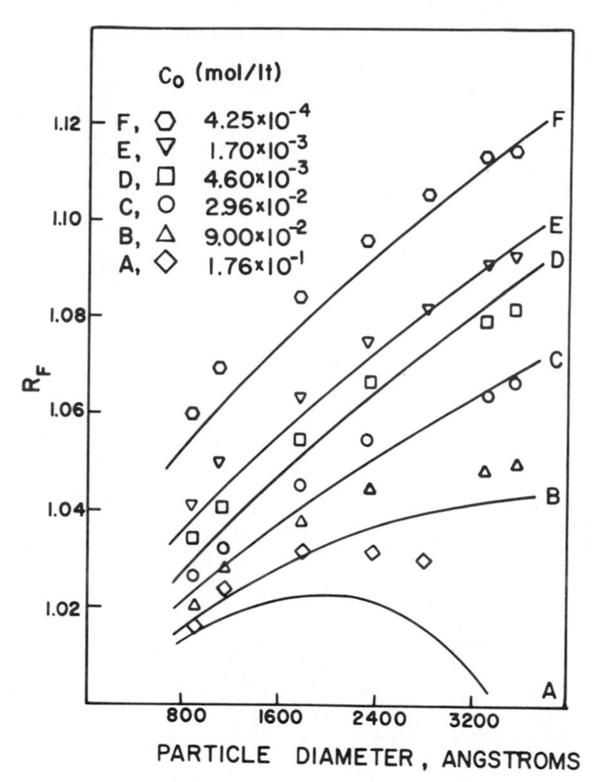

Figure 7. Comparison between calculated (solid line) $R_F - D_p$ relationship and data from Small (1) at various ionic strengths.

high ionic strength region by a slight upward adjustment of R_o and the Hamaker constant (15) however our purpose here is not to fit data but to demonstrate the efficacy of the capillary model in explaining the separation mechanism in HDC.

III. PARTICLE SIZE ANALYSIS: AXIAL DISPERSION CORRECTION AND METHODS FOR IMPROVED RESOLUTION

As a monodisperse colloid slug passes through the HDC columns, the particles will separate by axial dispersion. The turbidity output signal thus results as a bell-shaped curve rather than a single vertical line at the corresponding particle elution volume. The area under the monodisperse curve will depend on the number of particles injected and a size-dependent proportionality factor. For a polydisperse sample, the chromatogram will be a composite resulting from contributions of each individual particle population and will in general appear as a single broad peak in the detector output. In order to obtain accurate particle size distributions as well as meaningful averages, the output signal must be corrected for this axial dispersion (referred to as instrument spreading). The mathematical techniques for dispersion correction that can be used are essentially the same as developed for use in GPC and are broadly classified as Analytical and Numerical (16). The principal source of difference in the HDC correction methods arises from the extreme sensitivity of the output turbidity signal to particle size.

For the low particle concentrations used in our experiments (less than 0.01% by wt) it can be demonstrated that multiple scattering will be insignificant (17). The output signal at a given elution volume F(V) will thus be given by the linear superposition of the individual particle species contributions, each corrected for dispersion. If the number of species is large, the output signal is given by the Fredholm integral equation of the first kind

$$F(V) = \int_{V_a}^{V_b} f(V, y)\, A(y)\, dy \qquad (6)$$

f(V, y) the normalized dispersion correction represents the contribution of a specie eluting at y and A(y) the area under the chromatogram due to this specie. A(y) is related to the number of particles of each species N(y) by

$$N(y) = \frac{A(y)}{K(y)} \qquad (7)$$

In the case of a turbidity output signal

$$K(y) = \frac{R(y)x}{2.303} \quad (8)$$

with, x, the optical path length and R(y) the particle scattering cross section which can be calculated from the Mie theory (4). Equations 6-8 have been checked against material balances on our columns for several particle sizes and found to agree within experimental error (8). From these expressions the normalized PSD, $N^*(y)$, can be obtained as

$$N^*(y) = \frac{A(y)}{mA_T\ K(y)\ D_p(y)} \quad (9)$$

Where m represents the slope of the calibration curve and A_T the total area under the chromatogram. Several different techniques have been evaluated for solving equation 6 (8). An iterative method which was found to work well will be summarized here. A more complete discussion of the various methods (Analytical and Numerical) will be subject of a future publication.

Briefly the iterative method starts with a first estimate of A(y) obtained from the polydisperse chromatogram assuming no dispersion, equation 6 is then evaluated from which the computed chromatogram $F^*(v)$ is obtained. From this, the distribution A(y) is then corrected at each integration point depending on the error between the computed and measured chromatogram. The correction factor has the following form which is a generalization of a technique reported earlier by Ishige, Lee and Hamielec (18).

$$A_i^{j+1} = A_i^j \prod_{k=-n}^{n} \left(\frac{F^*_{i+k}}{F_{i+k}} \right)^{\alpha k} \quad (10)$$

where j refers to the level of iteration, F^* is the computed chromatogram and F the actual chromatogram the αk are weighting coefficients taken from the actual contributions of the neighboring sizes. The number of symmetric terms about F_i(i.e. n) was chosen according to the spread in the chromatogram (8). Figure 8 shows the fit which results for a polydisperse polystyrene synthesized in these labs. The crossed lines are calculated points, the solid

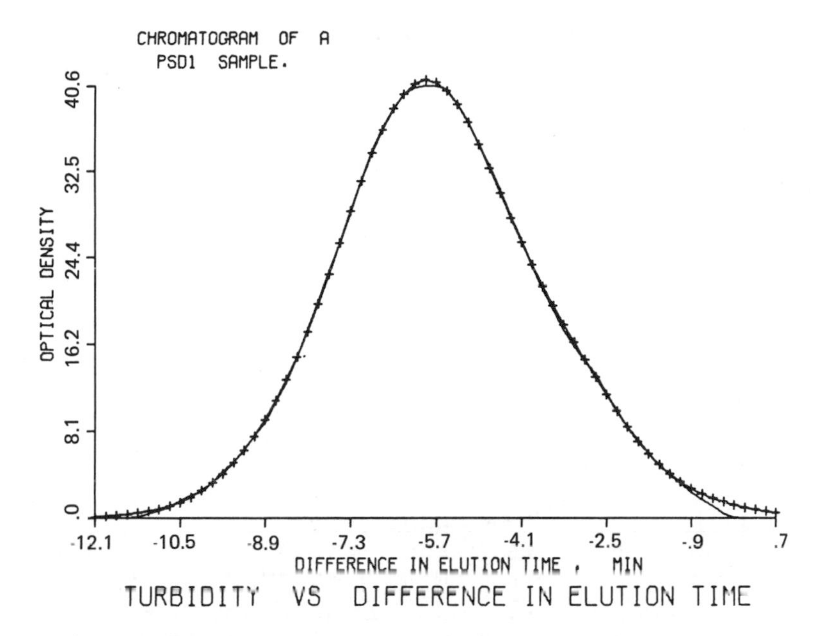

Figure 8. HDC Chromatogram of polydisperse polystyrene latex. Solid lines represent actual chromatogram, points represent calculated values.

line is the actual chromatogram (in this case n = 8). Figure 9 shows the continuous normalized distribution which results and the histogram represents the actual PSD as determined by electron microscopy. The fit is seen to be reasonably good specially for the larger particles. It is important to emphasize that the small mismatch in the chromatogram in the small particle range results in an appreciable error in the small particle region of Figure 9. Basically the error magnification results from the strong size dependence of K(y) in the small size range (essentially a 6th power dependence on D_p). This results in a very small signal representing a large number of particles and is more vividly demonstrated in Figures 10 and 11. Here we have a chromatogram for a 1 to 1, by number, mixture of 88 nm and 176 nm diameter particles where it can be seen that the smaller particle size shows only as a small shoulder in the output even though the particles are present in equal numbers. Figure 11 shows that the iterative method is able

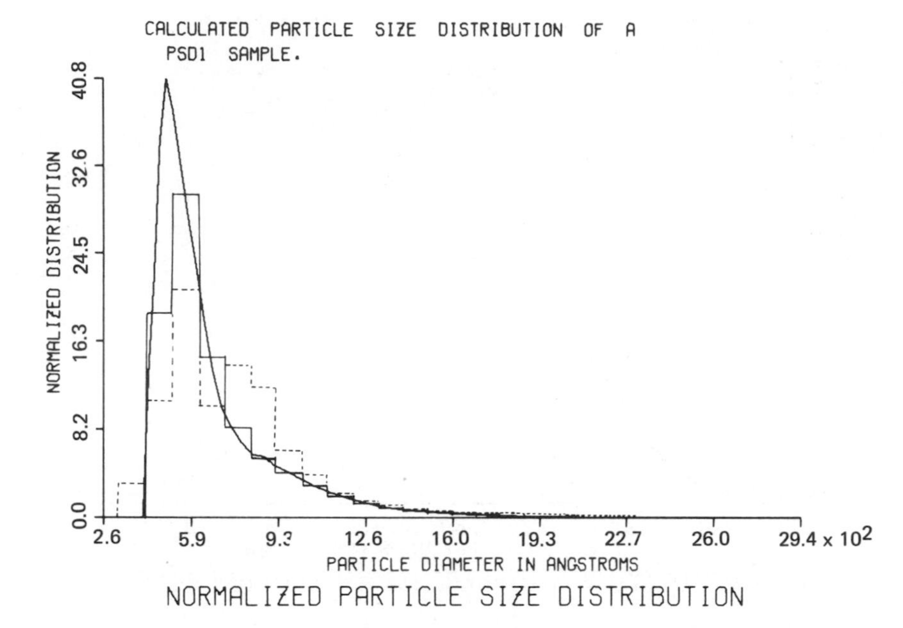

Figure 9. Particle size distribution resulting from calculations shown in Figure 8. Solid line represents calculated PSD histogram represents PSD from electron microscopy.

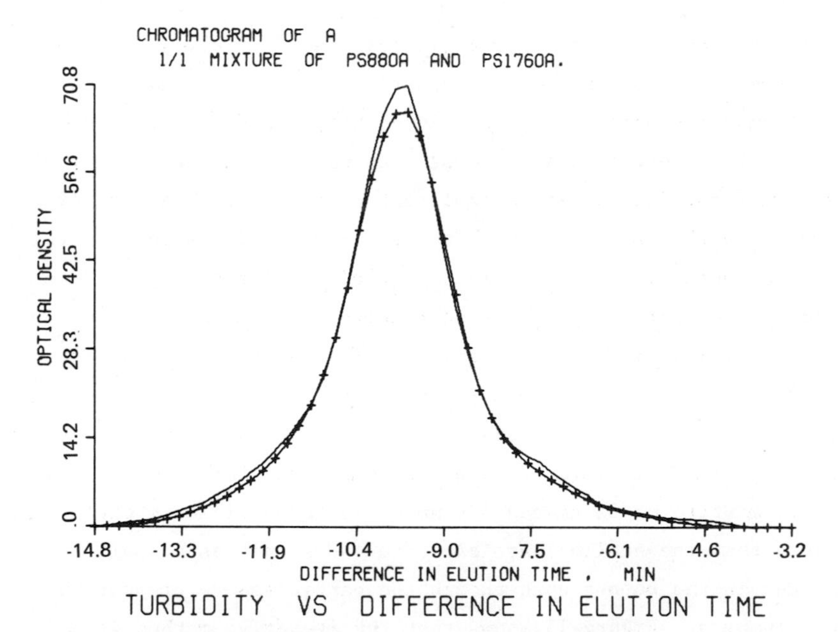

Figure 10. Chromatogram for 1/1 mixture (by numbers) of 88 and 176 nm polystyrene particles. Solid line actual chromatogram, points represent calculated values.

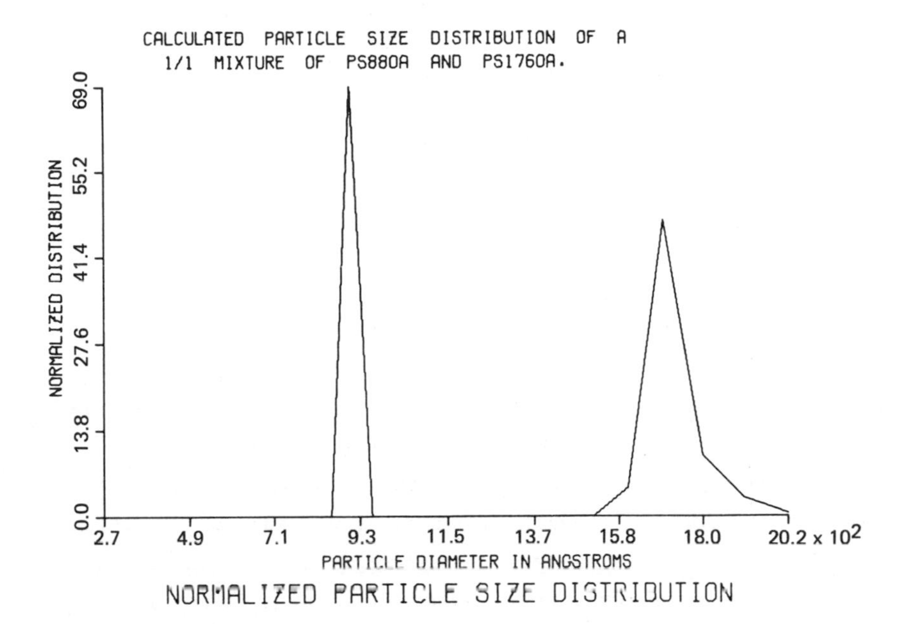

Figure 11. PSD for Figure 10, Ratio of Areas under 176 nm and 88 nm distributions is 1.5/1.

to predict the presence of the two populations although the number ratio of 176 to 88 has been changed to 1.5/1. Similar results were obtained with an integral method which assumes a non-Gaussian form for f(V, y) (8). Both methods, though they can be shown to lead to reasonably good results for the PSD, point to the need for improved resolution.

Resolution can be improved in two ways: 1) improved signal detection which would be less sensitive to D_p, and 2) porous column packing (6). In terms of signal detection one would be led to expect improvement by monitoring the change in refractive index where for dilute systems Zimm and Dandliker (19) have shown

$$\frac{dn^1}{dc} = \frac{3n_1 \; R_e(j\;)_{180^o}}{\alpha^3 \; \rho_2} \qquad (11)$$

with $\alpha = D_p/\pi\lambda$ being λ the wavelength of source in the medium and $R_e(j\;)_{180^o}$ represents the real part of $(j\;)_{180^o}$ which pertains to the light scattering in the forward direction as computed from Mie theory, n^1 is the refractive index of the dispersion and ρ_2 the density of the particles. In the limit for very small spheres the above expression reduces to

$$\frac{dn^1}{dc} = \frac{3n_1}{2\rho_2} \quad \frac{n_2^2 - n_1^2}{n_2^2 + 2n_1^2} \qquad (12)$$

which is the same result obtained from the Lorentz equation. Figure 12 illustrates graphically the dramatic improvements in signal intensity which result with the refractive index measurement. Here we have plotted calculations for the signal intensity for a series of polystyrene latex particle diameters relative to an equal number of 30 nm diameter particles. The refractive index signal was calculated from equation 11 which represents the K(y) term in equation 7 for differential refractometry. These calculations show a decrease in relative signal intensity of as much as 2 orders of magnitude in the small particle range indicating the possibility of peak separation for the chromatogram in Figure 10.

Figure 13 shows data from our labs of refractive index measurements taken on a Brice-Phoenix instrument. A series of monodisperse polystyrene latexes were used ranging from 88 to 357 nm in diameter. The wavelength of light was 546.1 nm hence the maximum value of α was 0.07. In this range one expects a linear realtionship between the refractive index difference between the colloidal suspension and the medium and the latex concentration (8) as is shown in these data. The scatter in the measurements is due most likely to the adsorbed surfactant and limited instrument resolution (the differential refractometer measured differences only to the fifth decimal place). The straight line through the data was computed from equation 12 and illustrates that HDC instrument resolution and dispersion analysis sensitivity will be greatly improved by using the refractive index signal (i.e. third power dependence on diameter for small particles versus sixth power for turbidity).

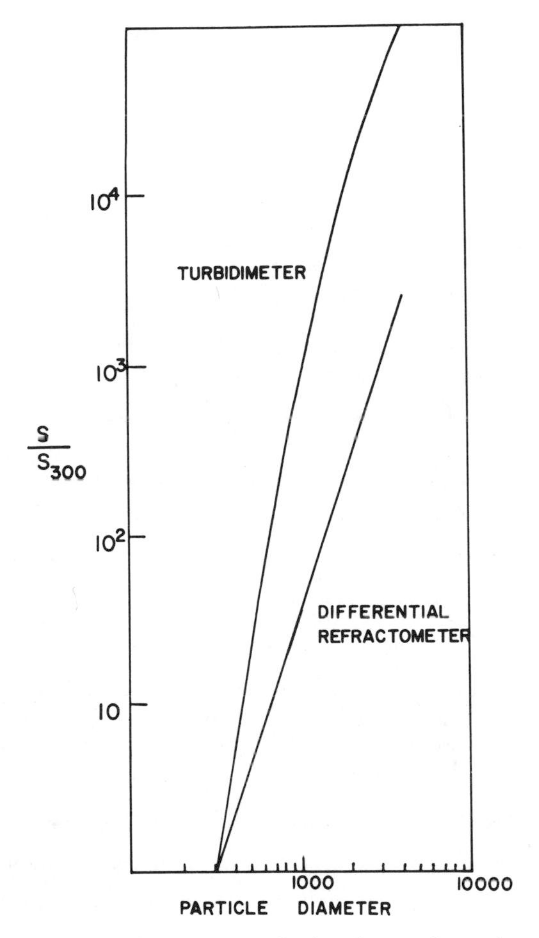

Figure 12. Relative intensity of the detection signal, for a turbidimeter and a differential refractometer, with respect to an equal number of particles 300 Å in diameter.

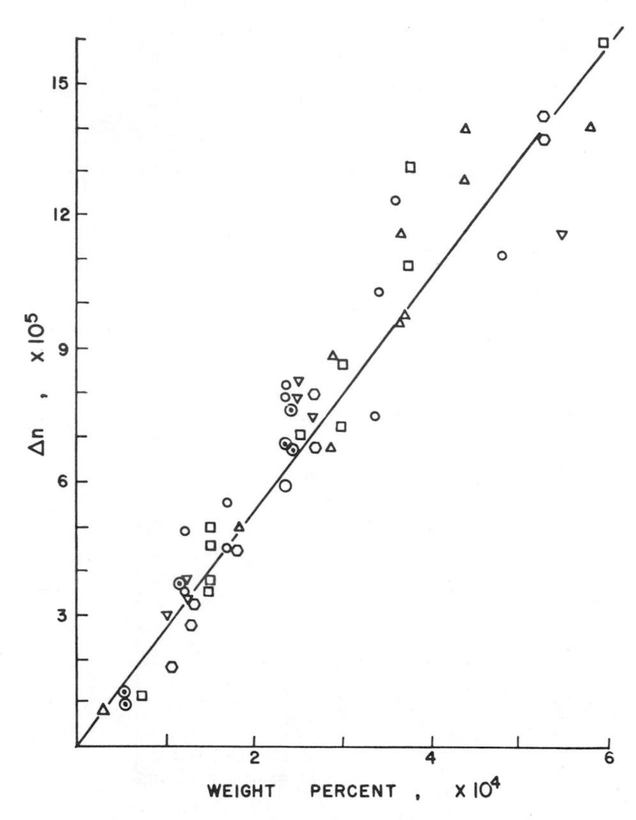

Figure 13. Difference in refractive index between polystyrene latex and medium versus particle concentration. Measurements made for a series of monodisperse latexes.

Later publications will deal with work currently being pursued in these labs utilizing porous packing systems to improve resolution. However, published data have already demonstrated improved resolution in terms of calibration curves with smaller slopes and increased values of the separation factor R_F (2,3,20).

REFERENCES

1. Small, H., J. Colloid Interface Sci., 48, 147 (1974).

2. Krebs, V. K. F., and Wunderlich, W., Angew Makromol. Chem., 20, 203 (1971).

3. Coll, H., and Fague, G. R., "Liquid Exclusion Chromatography of Colloidal Dispersions", paper presented at Cleveland - Akron GPC/LC Symposium, Cleveland, Ohio, April (1977).

4. Stoisits, R. F., Poehlein, G. W., and Vanderhoff, J. W., J. Colloid Interface Sci., 57, 337 (1976).

5. Small, H., Saunders, F. L., and Solc, J., Adv. Colloid Interface Sci., 6, 237 (1976).

6. McHugh, A. J., Silebi, C., Poehlein, G. W., and Vanderhoff, J. W., J. Collid Interface Sci., Vol IV, Hydrosols and Rheology, 549 (1976).

7. Scolere, J., MS Thesis, Lehigh University, Bethlehem, Pa. (1977).

8. Silebi, C. A., PhD Thesis in prep. Lehigh University.

9. Brenner, H., and Gaydos, L. J., J. Colloid Interface Sci., 58, 312 (1977).

10. Silebi, C. A., and McHugh, A. J., "An Analysis of Flow-Separation in Hydrodynamic Chromatography of Polymer Latexes", paper submitted to AIChE J.

11. Goldsmith, H. L., and Mason, S. G., J. Colloid Interface Sci., 17, 448 (1962).

12. Coldman, A. J., Cox, R. G., and Brenner, H., Chem. Eng. Sci., 22, 653 (1967).

13. Ottewill, R. H. and Walker T., Kolloid, Z.u.Z., 227, 108 (1968).

14. Small, H., "Hydrodynamic Chromatography - A New Approach to Particle Size Analysis", paper presented at Cleveland - Akron GPC/LC Symposium, Cleveland, Ohio, April (1977).

15. McHugh, A. J., Silebi, C., Poehlein, G. W. and Vanderhoff, J. W., "Hydrodynamic Chromatography (HDC) of Latex Particles", paper presented at A.I.Ch.E. Meeting, Houston, Texas, March (1977).

16. Tung, L. H., J. Appl. Polym. Sci., 10, 375 (1966).

17. Stoisits, R. F., MS Thesis, Lehigh University, Bethlehem, Pa. (1975).

18. Ishige, T., Lee, S. I., and Hamielic, A. E., J. Appl. Polym. Sci., 15, 1607 (1971).

19. Zimm, B. H., Dandliker, W., B., J. Phys. Chem. 58, 644 (1954).

20. McHugh, A. J., "Latex Particle Size Distributions by Chromatographic Methods", paper presented at Cleveland-Akron GPC/LC Symposium, Cleveland, Ohio, April (1977).

COMPARATIVE PARTICLE SIZE TECHNIQUES FOR POLY(VINYL CHLORIDE) AND OTHER LATICES

C. A. Daniels, S. A. McDonald and J. A. Davidson

Avon Lake Technical Center
BFGoodrich Chemical Division
Avon Lake, Ohio

Summary

Recently Davidson and Collins conclusively demonstrated that PVC latexes shrink about 20% in the beam of the transmission electron microscope (1). Furthermore, this work suggested that particles smaller than about 0.05 μm would not be seen at all. Since the measurement of PVC latex particle sizes below 0.2 μm and especially below 0.1 μm still represents a problem, a transmission electron microscope sample preparation technique employing vertical shadowing was developed (2). In this technique, the latex particles are shadowed using gold-palladium at an angle of 90° to the surface of the grid. Even though the particle shrinks in the electron beam, the circular shadow will still represent the original size.

Ten PVC latex samples in the range 0.05 μm to 1.0 μm were examined using this technique and the diameter found to be in good agreement with those obtained by other methods. The methods chosen for comparison were as follows: Fractional Creaming, Joyce Loebl Disc Centrifuge, Micromeritics Sedigraph, Optical Arrays and Coulter Counter. Care must be taken to recognize the affect of agglomeration if good agreement between various methods is to be obtained.

Other latices were also analyzed by these methods and, in addition, hydrodynamic chromatography. Comparative results for some polystyrene, polybutadiene and acrylic latices will be discussed.

I. INTRODUCTION

Recently, Davidson and Collins (1) have demonstrated that poly(vinyl chloride) latexes in the range 0.27-1.2 μm shrink when examined in the transmission electron microscope. These authors found relationships of the following type:

$$D_X = 1.4\ D_{EM} + .050 \qquad (1)$$

where D_X is the average diameter (μm) by the maximum-minimum light scattering technique at λ = 546 nm and D_{EM} is the observed electron microscope diameter (see Figure 1). The above relationship implies that particles below 0.050 μm will not be seen in the electron microscope under normal operating conditions. However, the inves-

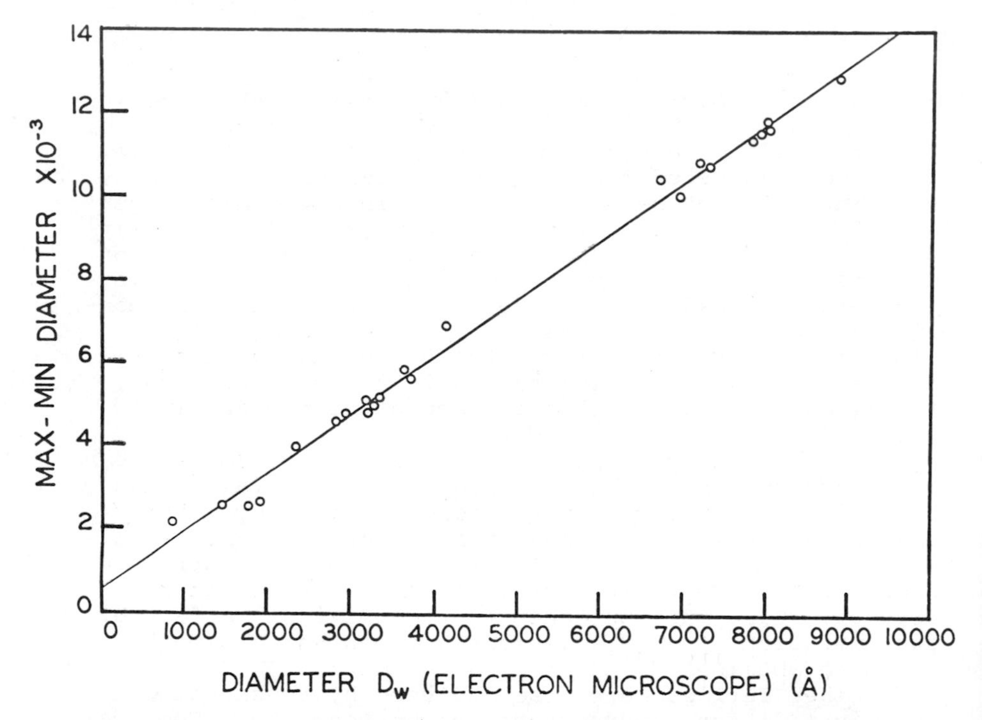

FIG. 1A. Weight average diameter (D_W) from electron microscopy versus the diameter determined by (max.-min.) light scattering (λ = 436 nm) for monodisperse PVC latices.

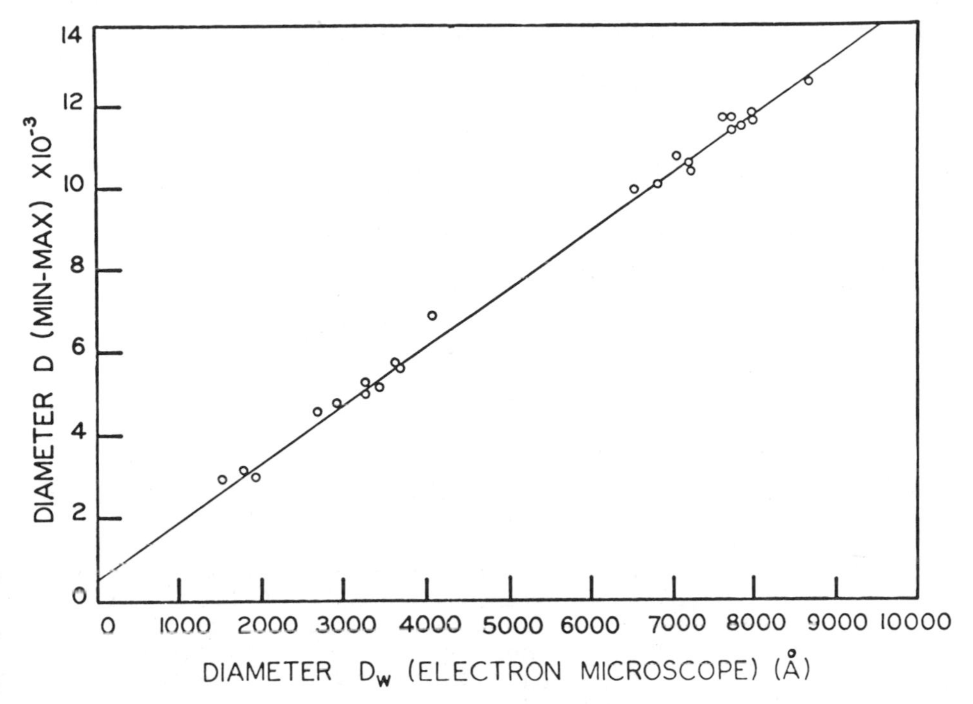

FIG. 1B. Weight average diameter (D_W) from electron microscopy versus the diameter determined by (max.-min.) light scattering (λ = 546 nm) for monodisperse PVC latices.

tigation of particles below 0.20 μm involves a number of experimental problems since few methods other than electron microscopy exist which enable particle size to be accurately determined in the range 0.03-0.20 μm.

In order to obtain accurate particle size data and to investigate the above mentioned range of particle size diameters using electron microscopy, a method of sample preparation using vertical shadowing prior to examination of the latex samples was developed(2).

In the previously mentioned investigation by Davidson and Collins (1), it was observed that shadowed PVC latex particles produced shadows which were larger than the corresponding particles. Furthermore, the size of these shadows at their widest point closely approximated the correct size of the original latex particles. By shadowing at 90° (vertical shadowing), the sizing of the shadows is much simplified since a circular rather than an elliptical shadow is produced. In addition, vertical shadowing enables a much denser field of particles to be examined, since the particle and shadow occupy the smallest possible area on the plate. In order to test the validity of this technique, comparisons were made with the Joyce Loebl Disc Centrifuge, Micromeritics Sedigraph, the Coulter Counter, Optical Array Measurements, Higher Order Tyndall Spectra (HOTS) and Fractional Creaming. These techniques were reviewed recently by Collins, Davidson and Daniels (3).

In order to extend these comparisons, other latex systems were investigated: polybutadiene, polystyrene and acrylic latices, both polydisperse and monodisperse in size distribution. In the case of poly(vinyl chloride), hydronamic chromatography was also employed as a first step in comparing this new technique with those previously employed.

II. MATERIALS

Seven poly(vinyl chloride) latexes prepared by emulsion polymerization in the range of 0.20-1.0 μm diameter were employed in this investigation. Two 20/80 methyl acrylate/PVC copolymer latexes were also investigated. These latter had particle diameters of about 0.05 and 0.08 μm, respectively.

The polystyrene monodisperse latices were obtained from Duke Standards. Polybutadiene latices were supplied by the BFGoodrich Chemical Division, as was the acrylic latex, which had a composition of about 85% ethyl acrylate and 15% by weight acrylic acid.

III. EXPERIMENTAL

Specimens for electron microscopy were initially prepared by evaporating a drop of dilute latex (10^{-3} - 10^{-6} g/cc) on a 200 mesh formvar coated copper electron microscope grid. At least six grids were prepared for each latex investigated. Six of the prepared grids were then placed on Scotch brand double-adhesive tape on a glass slide and placed in the Denton DV 502 vacuum evaporator.

The grids were positioned at 90° relative to the filament at a distance of about 13 cm. The filament employed was a 8.5 mm diameter tungsten basket fabricated from 0.051 cm wire (supplied by Ladd Research Industries). In it was placed a ~3 cm length of 0.020 cm 60:40 gold-palladium wire (supplied by Ladd Research

Industries). Following evacuation to at least 4 X 10^{-5} torr and degassing of the filament for ten seconds at about ten amperes current, the gold-palladium was evaporated by first raising the current to 20-30 amperes and observing the onset of "sparking" of the filament. When sparking first occurred, the current was momentarily raised to the maximum value (50 amperes) for about one second and then dropped to 20-30 amperes. From two to three of these short bursts of high current were required to completely evaporate the alloy. The experimental apparatus is shown in Plates I and II.

Electron microscope measurements were carried out on an RCA model EMU-3H using an accelerating voltage of 100KV. Electron micrographs were taken at the magnifications ranging from about 6,000X to 50,000X on 3-1/4 X 4 inch Estar base Kodak #4489 electron microscope film. Negatives were sized using a Bausch and Lomb 7X hand magnifier capable of reading to $\pm$ 0.05 mm. Magnification checks were made using a diffraction grating replica supplied by W. A. Ladd Company, Incorporated.

Optical array measurements were obtained using the technique employed by Kubitschek and others (3,4). Micrographs were taken on Eastman Kodak 35 mm Plus X film using a Zeiss Photomicroscope II equipped with a 100X phase contrast objective of NA = 1.25. The arrays were sized using a 10X microscope equipped with a filar micrometer capable of reading to 10 μm at the magnification employed. Magnification was measured by sizing a monodisperse Dow polystyrene latex, LS-063-A, using the diameter for this latex published earlier (1). The magnification on the film was 905X.

Higher Order Tyndall Spectra (HOTS) were run according to the procedures reviewed by Maron and Elder (5). Measurements using the Micromeritics Sedigraph (supplied by Micromeritics Instrument Corporation) were performed as recommended by the manufacturer, the samples having been diluted with triple distilled, "colloid free" water containing ~0.1 weight percent sodium lauryl sulfate surfactant. Similarly, the Joyce Loebl (supplied by the Joyce Loebl Division of Marco Scientific) Disc Centrifugation measurements were made utilizing the two layer sedimentation technique, wherein the latex was centrifuged through pure water with sodium lauryl sulfate as the surfactant. In order to eliminate interference by large outsized particles, especially in the very smallest diameter PVC samples, the diluted latex (~ 1%) to be injected into the disc was filtered through a Millipore type 0.2 μm filter prior to injection. This significantly reduced the percentage of large particles, and indicated that the agglomerates were, in fact, present in the sample and not caused by hydrodynamic steaming. Had steaming (shear or g-force induced agglomeration)

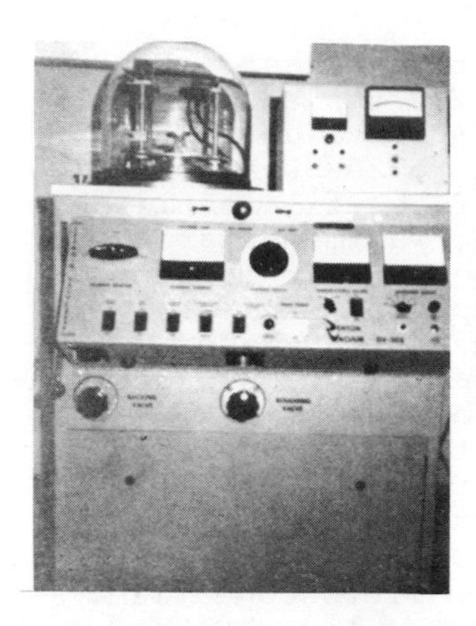

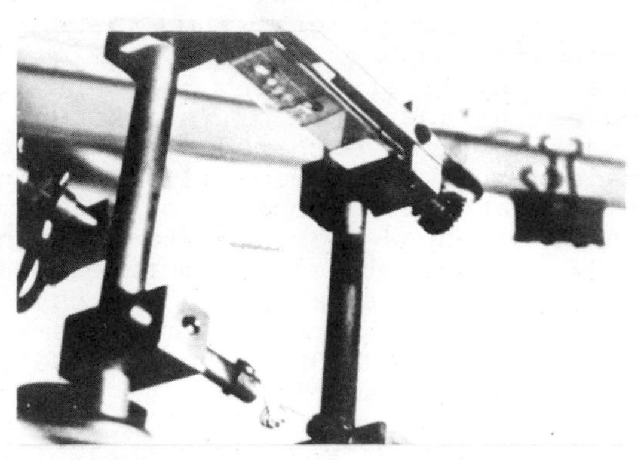

PLATE I. Experimental set-ups for vertical shadowing. Left, Denton DV502 vacuum evaporator set-up for vertical shadowing. Right, close-up of the Denton DV-502 evacuation chamber with the electron microscope grids 90° to the filament.

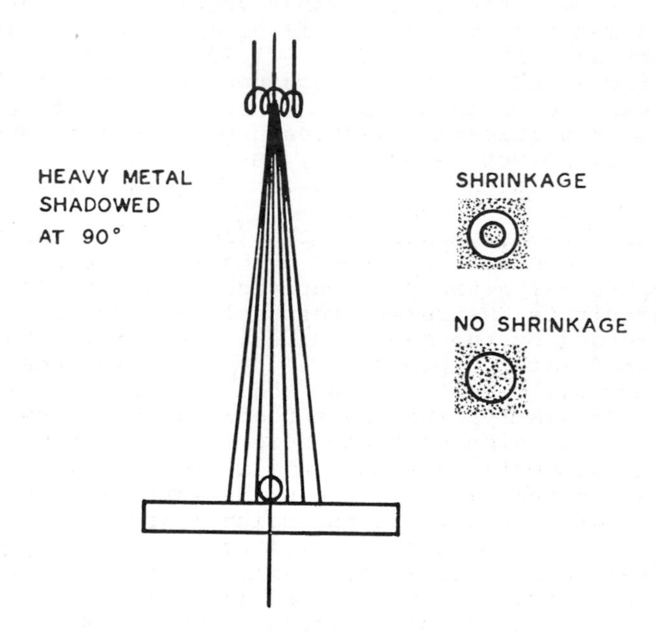

PLATE II. Illustration of observed shrinkage of particles in the electron beam and the relative dimensions of the particle and shadow produced.

been a factor, filtration would not have eliminated them. With larger latices, this step was eliminated.

Electrozone particle size analyses were run using the Model TA Coulter Counter, equipped with a 30 μm orifice (3). A 0.9% saline was used as the required electrolyte. Hydrodynamic chromatography results were kindly supplied by Professor G. Poehlein of Lehigh University on three polydisperse PVC latices. Only the median diameters were reported.

IV. RESULTS

Plate III shows a typical electron micrograph of a PVC latex after vertical shadowing. The original size of the latex particle is obtained by sizing the shadow outline. Table I presents a comparison of the number average diameter (D_n) as determined from both the particles and the shadows. A comparison of the average diameters obtained from Fractional Creaming, Joyce Loebl

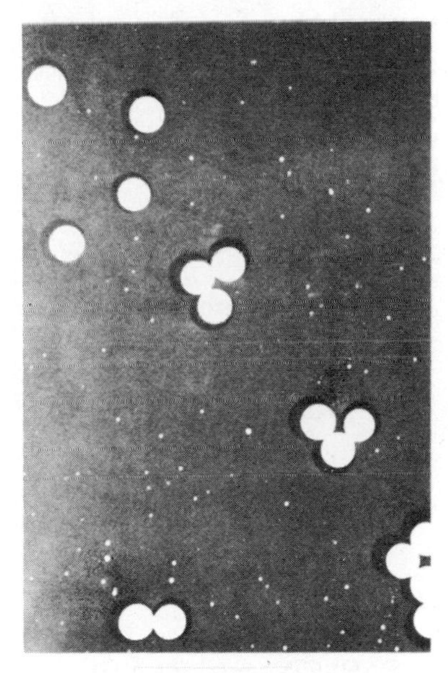

Sample B

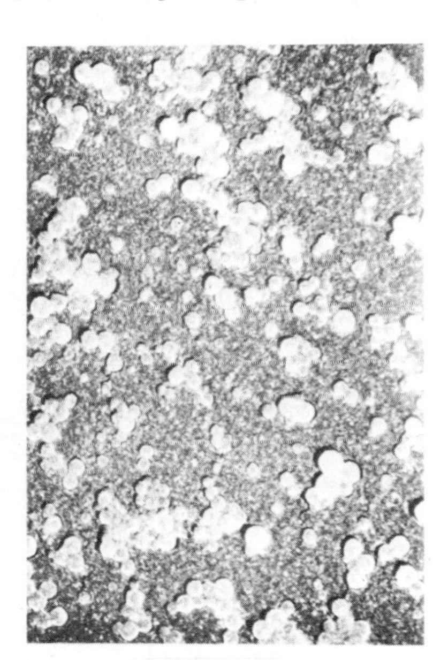

Sample I

PLATE III. Results of shrinkage of PVC in the electron microscope following preparation by vertical shadowing.

TABLE I

Comparison of Particle Size Diameters (μm) by Various Methods

PVC	TEM		Optical		Fractional	Joyce	Micromeritics	Coulter
Sample	Shadow	Particle	Array	HOTS	Creaming	Loebl	Sedigraph	Counter
mm	0.662	0.560	0.694 0.682	0.720	0.77			
A	0.976	0.816	1.03	1.04				1.09
B	0.938	0.773	0.996					
C	0.868		0.891	0.953				0.990
D	0.503		0.510	0.548		0.552	0.520	
E	0.484		0.491	0.512		0.543		
F	0.285			0.326		0.289	0.290	
G	0.200 0.198	0.161 0.166			0.195	0.215	0.187	
H	0.0822				0.0810	0.0780		
I	0.0525				0.0490	0.0780		
I		0.0610*						

* Particles were not shadowed.

Disc Centrifuge, Micromeritics Sedigraph, Optical Array, HOTS and Coulter Counter are also presented in the size range applicable to each technique (3). Note that the electron microscope shadow diameters are slightly smaller than those obtained by other techniques. However, the diameters determined by vertical shadowing are within 5% of those determined by other methods. In all cases the agreement between the various techniques is quite good. In the case of sample mm and samples A-F, the good agreement can be attributed to the nearly monodisperse character of these latexes. In this study, Optical Array diameters tended to run slightly high. The reason for this is, as yet, unexplained.

Further points can be made with respect to the data obtained from Sample I. Equation 1 implies that particles below 0.05 μm will not be seen in the electron beam. A method which detects and provides sizes for the small diameter particles will lower the average particle size dimeter relative to a method which misses the small particles. Sample I was also analyzed without prior vertical shadowing. This technique yields a particle size diameter of 0.0610 μm, which, when compared with the shadow diameter of 0.0525 μm, indicates that the smaller particles do disappear in the electron beam. At the same time, it has already been established that the particles do in fact shrink in the beam (1). Thus, we have concurrent effects which can result in both a decrease and an increase in the mean diameter (shrinkage causes the mean to decrease; small particles disappearing would result in a larger mean). Our data, showing the net result of these two phenomena, indicates that the "disappearance" effect outweighs the shrinkage. Similar data on other samples will have to be obtained to verify these contentions, and it is obvious that the distribution of sizes in the original sample (number of very small versus large) can have a dramatic effect on whether the mean diameter increases or decreases. The only obvious method for avoiding the possible confusion is to strictly utilize a vertical shadowing approach.

As one might expect, the largest variance in diameters between various techniques on a given sample occurred with the smallest latex, where a variation of about $\pm$ 6% was observed. For all other cases, variations range from 1-5%.

It should also be remembered that both Fractional Creaming and the Disc Centrifuge are working near the lower limit of their useable range (3) with the smallest latices used in this study.

Figures 2, 3 and 4 show particle size distribution data for Samples G, H and I, respectively. The plots are arithmetic probability plots of "weight percent less

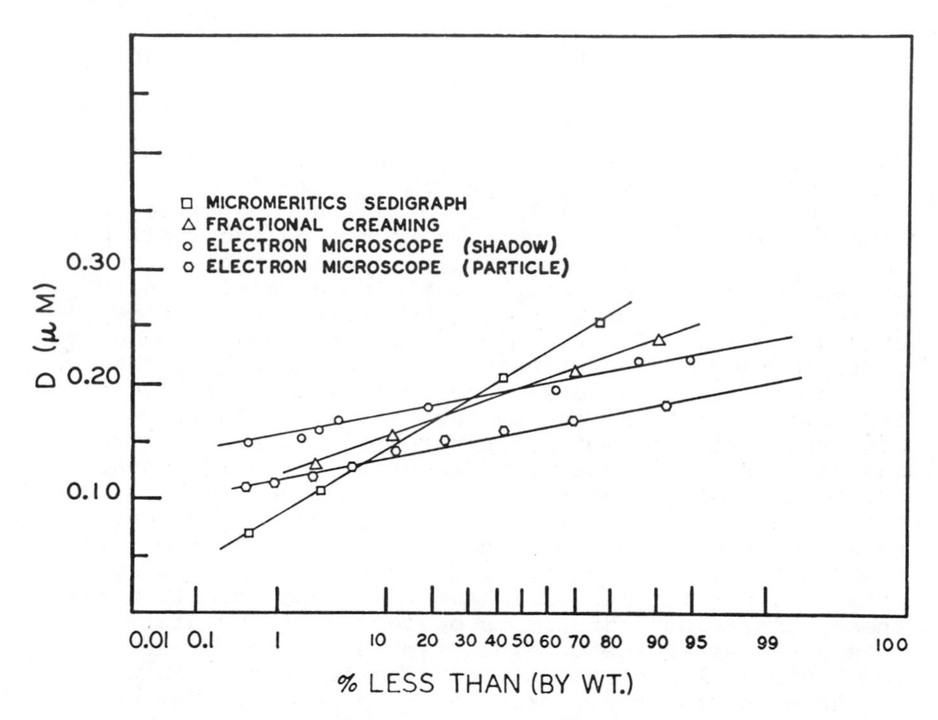

FIG. 2. Arithmetic probability plot of diameter (μm) versus cumulative percent less than by weight for sample G.

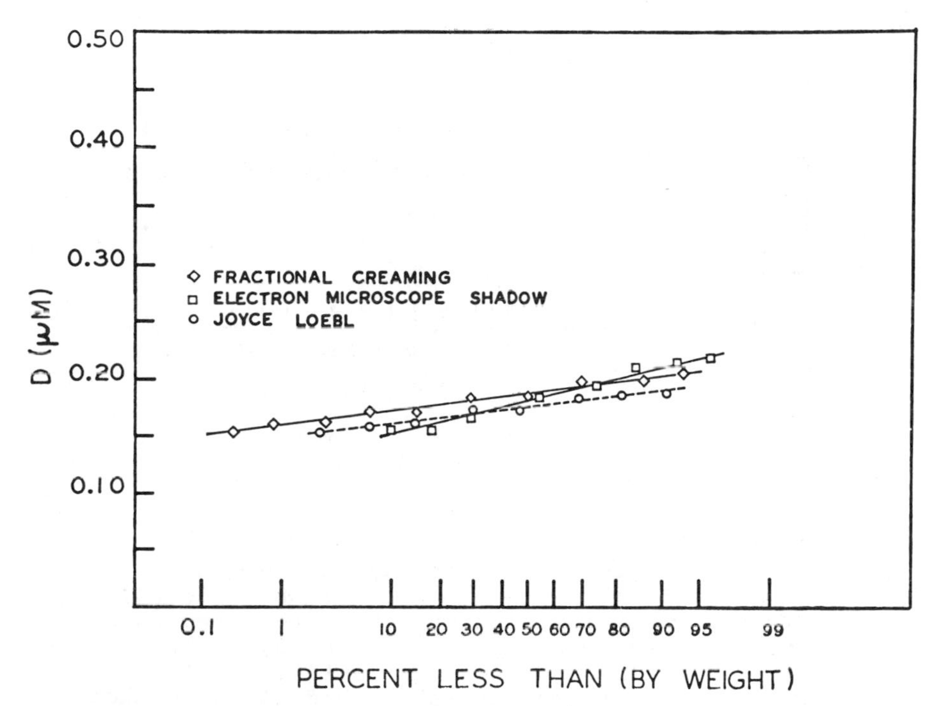

FIG. 3. Arithmetic probability plot of diameter (μm) versus cumulative percent less than by weight for sample H.

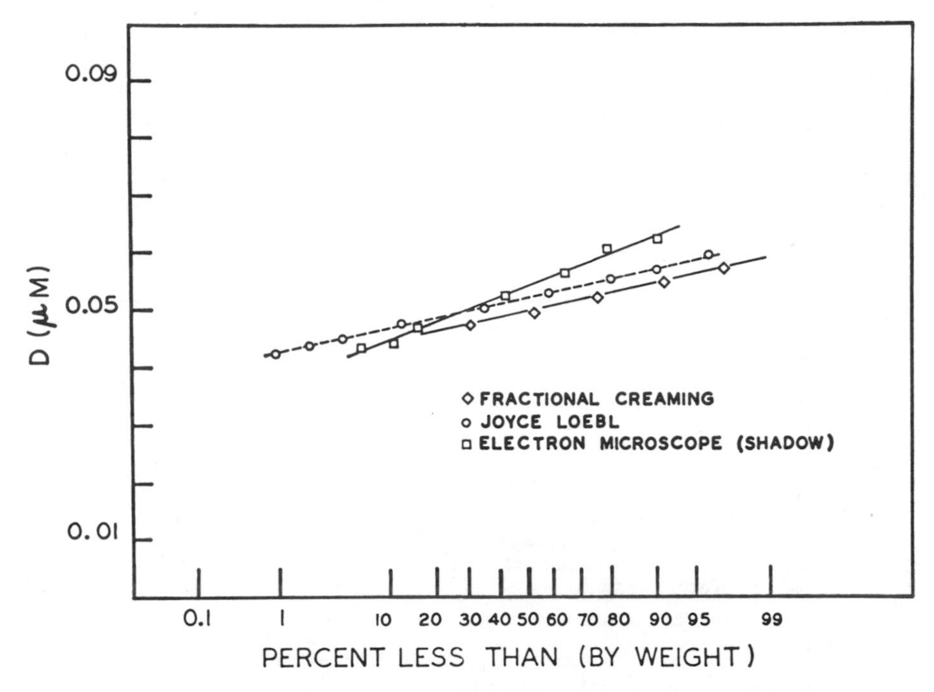

FIG. 4. Arithmetic probability plot of diameter (μm) versus cumulative percent less than by weight for sample I.

than" versus diameter. The straight lines obtained for all methods on Sample G (see Figure 2) indicate that this latex has a normal distribution. Note that two parallel lines are obtained for Sample G for the diameters determined from particles and their respective shadows by electron microscopy. This result would indicate that the correlation of the type obtained for PVC by Davidson and Collins (see Equation 1)(1) should hold at least down to 0.10 μm. Note that both fractional creaming and the Micromeritics Sedigraph indicate a broader distribution than the electron microscope. This may be due to the sensitivity of these methods to agglomerate size rather than primary particle size (6).

For Sample H with a mean diameter of about 0.08 μm, all three methods are in good agreement. Sample I was found to contain a few large particles above 0.10 μm, which strongly prejudiced the weight percent data obtained from electron microscopy. When all particles greater than 0.10 μm were dropped from the calculation of the distribution, a plot shown in Figure 4 was obtained with reasonable agreement in all cases.

A. Monodisperse Polystyrene Latices

Having properly calibrated the Joyce Loebl disc centrifuge (DCF) by obtaining the extinction coefficients for standard latices, comparisons were made between DCF results and those from electron microscopy on monodisperse polystyrene latex. Table II shows the data from both methods, and the agreement is excellent. Although the range of sizes reported in the table is rather limited, it demonstrates the utility of the DCF method. Moreover, very narrow size distributions for these latices by both methods were obtained. Mechanical instabilities in some DCF instruments can cause these polystyrene latices to appear polydisperse, but this is strictly an artifact of the technique when poorly balanced rotors are employed. Also, it is apparent that the agreement between the two methods is best at the largest diameter latex, and progressively poorer as the diameters become smaller. This is attributable to the inaccuracies in the DCF technique when the particle diameters are small, coupled with small differences between the latex particle density and that of the centrifugation medium, in this case, water. Interference by Brownian motion may also occur to hinder settling.

TABLE II

Comparison of Diameters
of Monodisperse Polystyrene Latices
by Disc Centrifugation and Electron Microscopy

D (μm), E.M.	D (μm), DCF
0.126	0.109
0.557	0.546
0.796	0.787

TABLE III

Comparison of Disc Centrifugation (DCF) and Fractional Creaming (FC) for Polybutadiene Homopolymer Latices

Sample		Cumulative Wt.% Below								
		0.15μm	0.20μm	0.26μm	0.30μm	0.36μm	0.40μm	0.50μm	D_{50} (μm)	Δ (μm)
A	FC	1.85	12.3	26.5	37.6	48.0	57.4	75.0	0.355	0.135
	DCF	1.68	12.0	29.0	41.8	55.0	64.0	80.5	0.385	0.120
B	FC	11.7	19.0	41.0	61.0	84.0	95.0	97.0	0.274	0.087
	DCF	4.0	18.5	42.0	58.5	77.0	88.0	97.2	0.279	0.085
C	FC	0.03	8.59	23.8	48.2	71.3	82.9	99.1	0.315	0.087
	DCF	0.46	12.0	33.0	49.0	67.0	77.0	90.1	0.304	0.094
D	FC	7.81	37.0	78.1	95.2	100	100	100	0.220	0.054
	DCF	14.4	41.0	72.8	87.0	95.4	98.4	99.9	0.214	0.061

B. Polydisperse Polybutadiene Latices

Four polydisperse polybutadiene homopolymer latices were examined by DCF and Fractional Creaming methods, and the comparison of distribution data is given in Table III. Here, the weight median diameters D_{50} vary only from ~ 0.20 to $0.35\ \mu m$; but the distribution breadths (measured by $\Delta = D_{50} - D_{16}$) range from ~ 0.05 to $0.13\ \mu m$. The latices for Joyce Loebl DCF measurements were hardened according to the procedure outlined by Bradford and Vanderhoff (7) and the diameters corrected accordingly. As shown in the table, the values for the distribution breadths, median and incremental percents for both methods are in excellent agreement. This demonstrates the utility of the DCF method for polybutadiene latices, where fractional creaming methods are accurate, but extremely tedious and time consuming.

Comparisons were also made on a broader spectrum of sizes of polybutadiene latices, using soap titration and DCF measurements. Since the soap titration method gives a surface average measurement (3), the data from the DCF were converted to these averages by standard methods. The range of diameters covered in this examination ranged from about 0.1 to $0.37\ \mu m$, and the data are given in Table IV. In the comparison, the values from the DCF tend to be slightly lower than that from the soap titra-

TABLE IV

Comparison of Diameters
by Soap Titration
and Disc Centrifugation Techniques
on Polybutadiene Homopolymer Latices

Sample	D_{sv} (Soap Titration) μm	D_{sv} (Calc. from DCF) μm
A	0.114	0.0850
B	0.214	0.199
C	0.371	0.341
D	0.142	0.141

TABLE V

Poly(vinyl Chloride) Latices Analyzed by Fractional Creaming, Disc Centrifugation and Hydrodynamic Chromatography (HDC)

Sample	D_{50}, FC, μm	D_{50}, DCF, μm	D_{50}, HDC, μm
1	0.0827	0.124	0.107
2	0.0750	0.107	0.0870
3	0.0705	0.113	0.092
4	0.112	0.166	0.145

tion. Two possible causes of this are the overcorrection of D_{SV} for distribution breadth, and/or the sensitivity of soap titration method to the presence of free soap in the aqueous layer. The reader is referred to reference 3 for a further description of the latter problem. It was not determined which of these problems contributed to the slight differences observed here.

C. Other Latices-Newer Methods

Finally, two additional comparisons were included in this study to illustrate where further studies need to be made. Four polydisperse poly(vinyl chloride) latices were each analyzed by FractionalCreaming, DCF and hydrodynamic chromatography techniques. In these analyses, since the hydrodynamic chromatography method had not yet been developed to the point of yielding distribution data, only median diameters, D_{50}, are compared in Table V. All of the latices were relatively small, in the 0.10 μm range. The DCF method, operating near the range where its accuracy is most affected by stability and other factors, gave the largest median diameters. Hydrodynamic chromatography diameters were half way between DCF and FC results. No explanation for the differences are obvious, since complete distribution data were not available from all methods. This system is still under investigation.

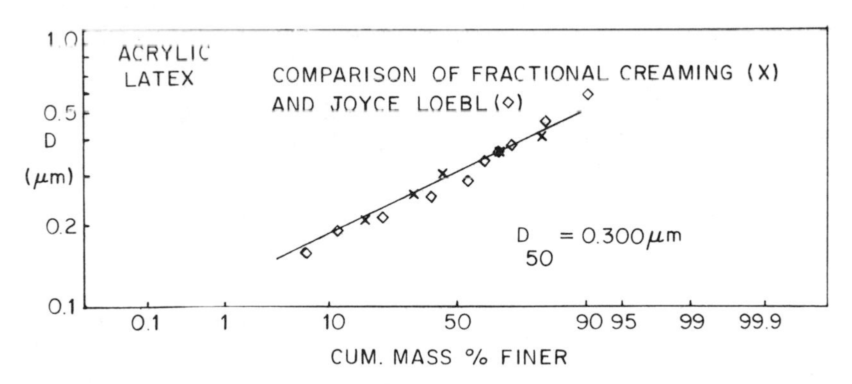

FIG. 5. Log probability plot of diameter (μm) versus cumulative percent less than by weight for an acrylic latex by fractional creaming and disc centrifugation.

Only one acrylic latex was included in this study, to provide another system where particle-media density differences might affect DCF results. As can be seen in Figure 5, the Fractional Creaming and DCF results are quite comparable for this latex of about 0.30 μm median diameter. Due to the Tg of this polymer, no electron microscopy evaluation could be accomplished.

V. CONCLUSION

This work demonstrates the utility of vertical shadowing in determining the size of PVC latices. This is especially to be recommended in the determination of latexes with diameter less than 0.10 μm. If comparisons of electron microscope measurements are to be made, care should be taken that the problem of agglomeration is carefully considered in the interpretation of the results. Furthermore, for a variety of latices, it can be concluded that Disc Centrifugation methods can also be used for both monodisperse and polydisperse latices provided a large enough density difference between particle and medium can be obtained, and provided that the size range does not extend significantly below about 0.070 to 0.080 μm. In the lower region, unless precautions regarding instrument sensitivity and stability are recognized, significant differences between DCF and other accepted methods may become apparent.

VI. ACKNOWLEDGMENT

The authors would like to thank the BFGoodrich Chemical Division for permission to publish this report.

VII. REFERENCES

(1) J. A. Davidson and E. A. Collins, Journal of Colloid Science, 40(3), 437 (1972).

(2) S. A. McDonald, C. A. Daniels and J. A. Davidson, Journal of Colloid and Interface Science, 59(2), 342 (1977).

(3) E. A. Collins, J. A. Davidson and C. A. Daniels, J. Paint Technol., 47(604), 35 (1975).

(4) H. E. Kubitscheck, "Ultrafine Particles", (W. E. Kuhn, ed.) p. 438, John Wiley (New York, 1963).

(5) S. H. Maron and M. S. Elder, Journal of Colloid Science, 18, 199 (1963).

(6) J. A. Davidson and H. S. Haller, Latex Particle Size by Dark Field, A Light Scattering Method, J. Colloid and Interface Sci., in press.

(7) E. B. Bradford and J. W. Vanderhoff, Journal of Colloid and Interface Science, 14, 543 (1959).

A. A. Bibeau (Union Carbide)

Have you taken into account the effect of extinction coefficient changes on disc centrifuge results?

Author We have experimentally determined the extinction coefficients for the various sizes of PVC latices used in the study. This was also done for polybutadiene latices, and less completely (in size range) for polystyrene latex. In acrylic systems, we have yet to complete this determination.

THE EFFECT OF ANGULAR RESOLUTION ON THE DETERMINATION OF PARTICLE SIZE DISTRIBUTION OF POLYMER LATEXES BY LIGHT SCATTERING

R. L. Rowell, J. W. Parsons, J. F. Ford and S. R. Vasconcellos, Department of Chemistry, University of Massachusetts, Amherst, Massachusetts

The early history of the measurement of absolute angular intensities for Mie scatterers has been reviewed by Kratohvil and Smart (1) who explored the use of polarized incident radiation and the effect of solid angle subtended by the receiver aperture. They concluded that a planar angular spread of $\pm 1^\circ$ did not influence the results significantly. The angular spread was computed from the dimensions of an aperture inserted immediately in front of the photomultiplier in its housing.

Extension of the measurement of absolute angular intensities to polydisperse systems of Mie scatterers was carried out by Rowell, Wallace and Kratohvil (2) who also neglected the effect of receiver aperture.

In recent work on the depolarization of water by Farinato and Rowell (3) it was shown that acceptance angle of the receiver was very important. Accordingly, we decided to reinvestigate the effect of acceptance angle for Mie scatterers.

Typically, the simplest collimating system employs two slits separated by a distance d. In such a case it is not sufficient to characterize the angular resolution by a single constant. An improved characterization of photometer geometry involves acceptance angles ϕ and ψ and a simple slit function that has been given elsewhere (3).

The effect of finite acceptance angle is to integrate over the source functions used in the light scattering. Experimentally, acceptance angle integration gives the same kind of result as a finite polydispersity i.e. erosion of the maxima and filling-in of the minima.

In the present work we investigate the effect of acceptance angle and express the results in terms of an apparent polydispersity. It is shown that for small acceptance angles, the slit-width integration amounts to $2.14 \times 10^{-3}\ \sigma_0$/deg where σ_0 is the zeroth-order log normal standard deviation and deg refers to the angle ψ as described elsewhere (3). The reported coefficient applies to α_M of 10 and a relative refractive index m of 1.2. In this domain, a 2° angular spread would give a σ_0 of 0.005 which is a large part of the apparent polydispersity of the latex.

REFERENCES

1. J. P. Kratohvil and C. Smart, J. Colloid Sci. 20, 875 (1965).
2. R. L. Rowell, T. P. Wallace and J. P. Kratohvil, J. Colloid Interface Sci. 26, 494 (1968).
3. R. S. Farinato and R. L. Rowell, J. Chem. Phys. 65, 593 (1976).

Discussion

M. N. YUDENFREUND (Drew Chemical Corp.): Your high resolution technique appears to be amenable to a determination of distribution breadth of the latices discussed previously by Dr. Daniels. This would be a good way of resolving the problems encountered with disc centrifugation techniques.

PROFESSOR ROWELL: Yes, in other work which is in preparation, we have obtained good agreement between our measurements and measurements by electron microscopy, quasi-elastic light scattering and angular light scattering on an aerosol of latex particles.

C. C. GRAVATT (National Bureau of Standards): Have you looked into the computer programs for similar slit corrections by users of SAXS and SANS methods?

PROFESSOR ROWELL: We know these are available and that they do apply to our situation. At this time, we are merely pointing out the importance of the slit correction in angular light scattering since it has apparently not been considered a significant error.

JOHN W. VANDERHOFF (Lehigh University): Would you care to comment on the effect of distribution of wavelengths in the Brice-Phoenix and in your instrument?

PROFESSOR ROWELL: We have not measured the effect but we would expect it to be small.

EMULSIONS AND DISPERSIONS

NEW MICROEMULSIONS USING NONIONIC SURFACTANTS AND SMALL AMOUNTS OF IONIC SURFACTANTS

Gunilla Gillberg[*], Leif Eriksson[**] and Stig Friberg[*]

[*]Department of Chemistry
University of Missouri-Rolla
Rolla, Missouri

[**]The Swedish Institute for Surface Chemistry
Stockholm, Sweden

Abstract

The influence of small additions of ionic surfactant to a nondisperse nonionic surfactant on the microemulsions formed in the system water-hexadecane-surfactants was studied. The investigation showed an expected rise in the temperature at which maximum solubility was observed (the HLB-temperature) but also a pronounced extension of the solubility area to lower surfactant concentrations than in the case of the pure nonionic surfactant. In addition the temperature stability of these microemulsions improved. The addition of ampholytic surfactant at its isoelectric point as an internal quaternary ammonium surfactant gave no influence on the solubility area at these low concentrations. A rational explanation of these effects was found in the repulsion between the surfactant ions is enough to counteract the binding energy which would cause a transition from the surfactant phase to an inverse micellar phase.

Microemulsions today attract a large technological (1-2) and scientific interest. So far the most investigated microemulsions are the W/O microemulsions formed in systems of an ionic surfactant, a cosurfactant, an oil and water. Different opinions of their nature and stability have been introduced ranging from an approach limiting the basis for stability to one factor, a negative interfacial tension (3) to more realistic treatments (4-5) including four terms of the total free energy. The W/O microemulsions with these components are formed by addition of hydrocarbon to the inverse micellar solutions formed by the three structure-forming components water, ionic surfactant and cosurfactant (4-8). These W/O microemulsions are accordingly thermodynamically stable colloidal solutions and in general show stability over a large temperature range. A disadvantage from a technical point of view of the W/O microemulsions is the need of comparatively high amounts of surfactant and cosurfactant to solubilize the water (of the magnitude 1 part of surfactant + cosurfactant for 2-3 parts of water).

Microemulsions in which simultaneously large amounts of water and oil may be solubilized by small amounts of surfactant are found in the "surfactant phase" most commonly observed in systems of nonionic surfactant, oil and water. A surfactant concentration of 6-8% is sufficient in order to solubilize equal amounts of oil and water using monodisperse nonionic surfactants. This surfactant phase was first observed by Shinoda, et al., (9-12) in the range of PIT (Phase Inversion Temperature) or the HLB-temperature. It was later (13-16) shown to be a separate solubility region with a minimum amount of surfactant needed at the HLB-temperature, below which it separates from a sectorial solubility region from the aqueous corner. These microemulsions with monodisperse surfactants have a small temperature range for stability (≈5 -10°C) which limits their use for most technical applications.

Shinoda and Kunieda (17) showed that admixtures of nonionic and ionic surfactants exhibited a larger temperature stability than the systems with the pure nonionic surfactants. However, these systems in which calcium dodecyl sulfate was added to a monodisperse tetraethylene glycol dodecyl ether in a weight ratio of 1:3 or higher led to enhanced surfactant concentrations (15) to the magnitude needed in microemulsions with ionic surfactants.

These investigations were concerned with large additions of an ionic surfactant of moderately hydrophilic character. The concentration of the ionic surfactant was of the same magnitude as the nonionic one. So far few investigations have been made on extremely small additions of a strongly hydrophilic surfactant to the nonionic surfactant. Such additions have earlier been shown to have a pronounced influence on the cloud point; additions of as little as .02% increased the cloud point by 35°C. Similar changes have been noticed in the presence of an organic solubilizate (12).

These results indicated a promising possibility to improve the temperature range for microemulsions with low emulsifier content and an evaluation of the phase behavior of systems of nonionic surfactants combined with minor amounts of a highly hydrophilic surfactant was considered of general interest.

EXPERIMENTAL

Materials: The nonionic surfactant was tetraethylene glycol dodecyl ether (TEGDE) (Nikkol). Gas chromatography showed that it contained some amounts of hexaethylene glycol dodecyl ether. The PIT of this delivery also showed to be higher than the surfactant used in earlier investigation (13-15). Sodium dodecyl sulfate ($NaC_{12}SO_4$) (BDH specially pure) was used without further purifications. Sodium laurate (NaLa) was prepared from dodecanoic acid and sodium ethylate in a nonaqueous alcohol solution (17). Dodecyl-β-alanine (C_{12}-A) was synthesized from n-dodecylamine (Kebo, puris) and β-propiolactone (Merk zur Synthese) (18). The sodium salt and HCl-salt were prepared by precipitation with sodium ethylate and conc. HCl respectively in ethanol solution and recrystallized once. Tetradecyl-β-propiobetaine (C_{14}-PB) was synthesized from N,N-dimethyl-n-tetradecylamine (K & K) and β-propiolactone (18). Dodecylammonium chloride ($C_{12}NH_3Cl$) was prepared from n-dodecylamine (KEBO, puris) and conc. HCl in ethanol. Sodium dodecyl ether sulfate ($C_{12}(EO)_3SO_4^-Na^+$) with an average 3 ethyleneoxide units per molecule was a gift from Berol Kemi AB. The hexadecane (Fluka p.a) was used without further purification. The water was twice distilled.

HLB-Temperature: Hexadecane and water in a weight ratio of 1 to 1 and surfactants in a total concentration of 5-20% were weighed into ampoules, which were then sealed. The samples were homogenized by vibration for a few seconds by means of a New Science NV, Vibromix PS. The samples were then placed in a thermostat, the temperature of which was raised in increments of 2.5°C in the temperature range of 15°C to 60°C. The storage time at each temperature was at least 30 minutes. The solubility areas were marked as the beginning and the end of the temperature range where the solutions became transparent.

Solubility Regions: The regions of the isotropic solutions in the systems of water-hexadecane-surfactants at a given temperature were determined by weighing two of the components into an ampoule to which the third component was added stepwise. After each addition the sample was vibrated for a few seconds and returned to the thermostat. The solubility area was marked as the beginning and end of the concentration range where the mixtures became transparent. A few samples close to the solubility boarders were sealed and stored for 48 hours.

RESULTS

The first part of this section will describe the influence of the presence of small amounts of an ionic surfactant to the nonionic system for the most pertinent W/O ratio of one. These results formed the basis for the selection of compositions and temperatures to be investigated in more detail. The results of these determinations of solubility areas at constant temperatures are given in the following section.

A. Isotropic solutions containing equal amounts of water and hexadecane.

Fig. 1A shows the temperature dependence for the solubility area for solutions containing water and hexadecane in a weight ratio of 1:1 and with various amounts of TEGDE. The minimum amount of surfactant needed to give an isotropic solution was 12w-% and the HLB-temperature for this system was 30°C. The earlier used TEDGE (References, 8-10) was of better purity and had an HLB-temperature of 25°C. It has earlier been shown (15) that surfactant phases with mixtures of different ethylene oxide chain lengths of the surfactant require higher amounts of surfactant and the present results are expected.

The effect of adding various small amounts of $NaC_{12}SO_4$ to the TEGDE is shown in Fig. 1B. The addition of the ionic surfactant in a ratio of 1:199 to the nonionic surfactant as expected increased the HLB-temperature but it also reduced the amount from 12 to 6% by weight of surfactants needed to give an isotropic solution. In addition to the reduced need of surfactants a comparison of Figs. 1A and B actually shows an improved temperature stability of the solutions close to these minima also. With the ratios $NaC_{12}SO_4$ TEGDE equal to 1:99 and 1:49 two solubility areas within different temperature ranges are observed arising from the usual trend of the sectorial solubility area from the water corner. One of those had an "HLB-temperature" of 35°C for both $NaC_{12}SO_4$-concentrations the other solubility region is shifted to higher temperature, with larger $NaC_{12}SO_4$-concentration.

For the $NaC_{12}SO_4$ TEGDE ratio equal to 1:49 the amount of surfactant needed to solubilize equal amounts of oil and water was high; similar to the original system, but with an enhanced temperature stability (10°C). The behavior is now approaching that of the high ratio system of Shinoda (7).

Fig. 1C shows the influence of the type of ionic surfactant added to the nonionic surfactant in the ratio 1:199. The addition

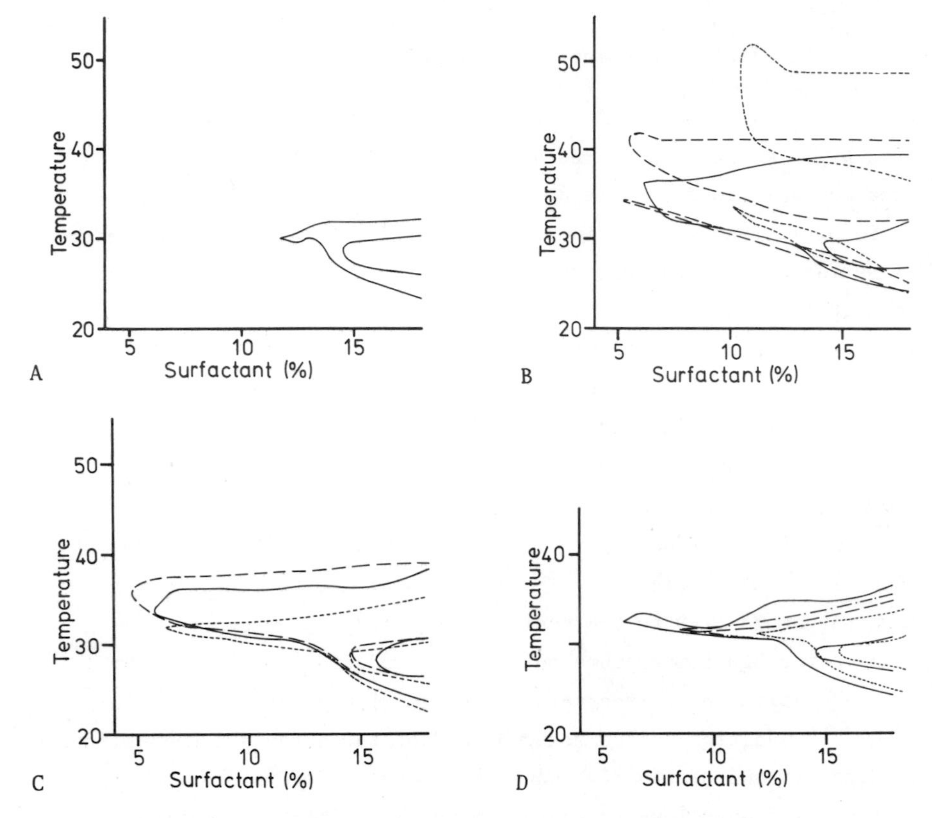

Fig. 1. Minimum surfactant requirements for obtaining isotropic solutions containing hexadecane and water in a weight ratio of 1:1 as a function of temperature.
TEGDE = tetra ethylene glycol dodecyl ether

A. Surfactant: TEGDE

B. Surfactant: TEGDE and sodium dodecyl sulfate in the weight ratio of 199:1 (——); (— — —); 49:1 (-----).

C. Surfactant: TEGDE and ionic surfactant in the weight ratio of 199:1 sodium laurate (——);
dodecyl ammonium chloride (— — —);
sodium dodecyl ether sulfate (-----)

D. Surfactant: TEGDE and ampholytic surfactant in the weight ratio of 199:1 dodecyl alanine hydrochloride (———);
dodecyl alanine hydrochloride and dodecyl alanine in weight ratio of 1:4 (— — —);
dodecyl alanine (----);
sodium dodecyl alanine (—.—.)

of $NaC_{12}(EO)_3SO_4$ to the TEGDE changed the HLB-temperature only with 2°C, but the minimum surfactant needed to solubilize equal amounts of water and hexadecane was reduced from 12 to 6% by weight. Addition of sodium laurate to TEGDE gave microemulsions at slightly lower surfactant concentration and with improved temperature stability. The mixture of TEGDE and dodecyl ammonium chloride required the lowest amounts of surfactant.

The ionization degree can be varied at a constant amount of added surfactant to the TEGDE by the use of ampholytic surfactant additive while retaining the basic structure of the surfactant. The result of such determinations is given in Figure 1D, with the ampholytic surfactant dodecyl-β-alanine ($C_{12}A = C_{12}H_{25}\text{-}NH_2\text{-}CH_2\text{-}CH_2\text{-}COOH$). The addition of 0.5% $C_{12}A$ to TEGDE had practically no effect on the region of existence of the isotropic solution (compare Fig. 1A with Fig. 1D, curve -----). Addition of the acidic salt $C_{12}A^+Cl^-$ ($C_{12}H_{25}\text{-}\overset{+}{N}H_3\text{-}CH_2\text{-}CH_2\text{-}COOH$) gave a slight increase in the HLB-temperature but also meant that much less surfactant was needed in order to give the isotropic solutions (Fig. 1D, curve ————). Addition of the same amount of the basic salt $C_{12}A^-Na^+$ ($C_{12}H_{25}\text{-}NH_2\text{-}CH_2\text{-}CH_2\text{-}COO^-Na^+$) did not have the same strong influence (Fig. 1D, curve —.—.—). This addition was more comparative to the addition of 0.5% of a mixture of $C_{12}A^+Cl^-$ and $C_{12}A$ in the weight ratio of 1:4, pH 3.8 (Fig. 1D, curve — — —). The corresponding mixture of $C_{12}A^-Na^+$ and $C_{12}A$ (weight ratio 1:4) had no effect on the solubility region when added in a concentration of 0.5% to the TEGDE. The same was true for the 0.5% addition of an internally ionized dodecyl-β-propiobetaine ($C_{12}H_{25}\text{-}\overset{+}{N}(CH_3)_2\text{-}CH_2\text{-}CH_2\text{-}COO^-$).

B. Solubility areas at constant temperature.

The TEGDE used contained some amounts of hexaethylene glycol dodecyl ethers and the solubility regions for isotropic solutions were redetermined at 25, 30 and 35°C. Fig. 2 shows an HLB-temperature of 30°C. The corresponding results for a surfactant

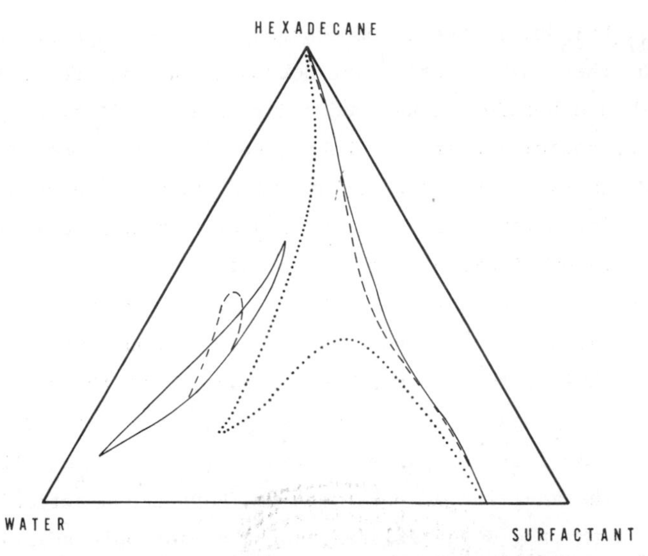

Fig. 2. Regions of isotropic liquids in a system of water, hexadecane and a surfactant (tetraethylene glycol dodecyl ether).
Temperature: 25°C (——);
30°C (– – –);
35°C (.....).

consisting of a mixture of TEGDE and $Na^+C_{12}SO_4^-$ in the weight ratio of 199 to 1 are given in Figs. 3A and 3B. The influence of the minute amounts of ionic surfactant was pronounced. The isotropic liquid solution sector (Fig. 3A) extending from (20, 25, 30°C) the water corner was observed in the temperature range 20-30°C. The stability of the two small isotropic islands observed at 25 and 30°C and of the L_2 area to the right in the diagrams have not been checked by long time storage. The usual isolated island (13) was not observed at the HLB-temperature (Fig. 3B curve, –.–.—); instead a large solubility region coalesced with the L_2 area was found.

This solubility region appeared of unusual shape and would actually permit the preparation of microemulsions with a wide range of oil/water ratios using only small amounts of surfactant. An

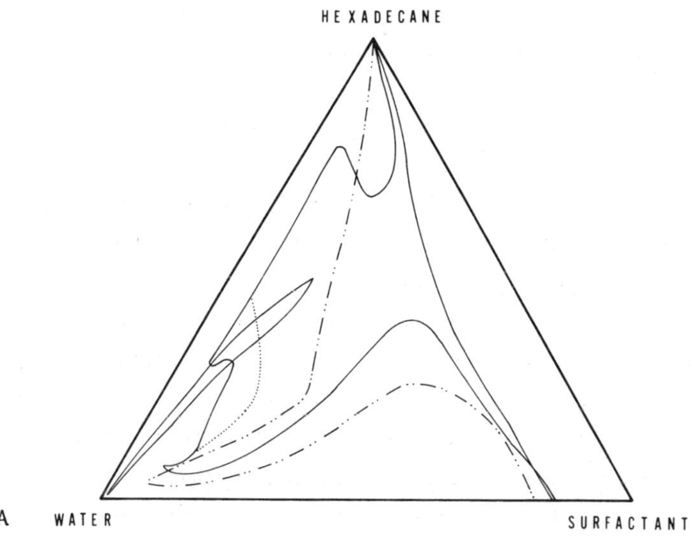

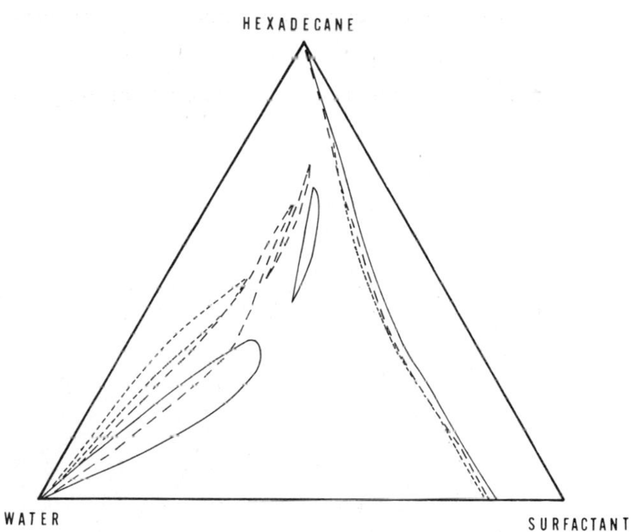

Fig. 3. Regions of isotropic liquids in a system of water, hexadecane and a surfactant (tetra ethylene glycol dodecyl ether and sodium dodecyl sulfate in a weight ratio of 199:1)

A. Temperature: 20°C (——);
25°C (— — —);
30°C (-----).

B. Temperature: 30°C (-----);
35°C (—.—.);
40°C (—..—..—).

addition of only 6% of the surfactants would be sufficient to prepare isotropic solutions with weight ratios of water to hexadecane ranging from 65:13 to 20:80. In the far left corner of the isotropic region (left of the dotted line) the solutions showed anisotropy when sheared. Addition of water in excess of the phase limit to these solutions caused a separation into two isotropic solutions and a liquid crystalline phase.

At 40°C no pronounced effects of the added anionic surfactant were observed (Fig. 3B, curve —..—..—); the solubility region attained an expected shape (13). At an addition of the cationic surfactant, dodecyl ammonium chloride to 0.5% to the TEGDE an identically large microemulsion region was obtained at 35°C (compare Figs. 3B and 4). Anisotropy at shear was an expected increase of 5°C from the more monodisperse surfactant earlier used (13-15) also observed in the same region. This fact agrees well with the general behavior of the corresponding temperature diagrams (Figs. 1B and 1D).

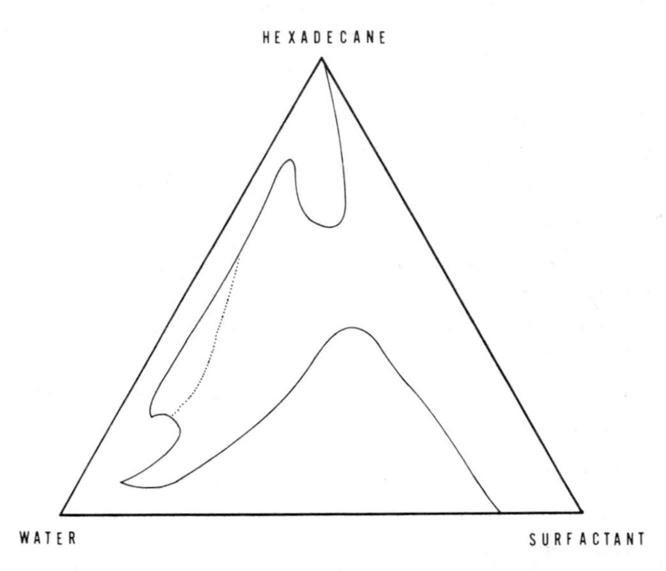

Fig. 4. Regions of isotropic liquids in a system of water, hexadecane and a surfactant (tetra ethylene glycol dodecyl ether and dodecyl ammonium chloride in a weight ratio of 199:1) at 35°C.

DISCUSSION

The present results demonstrated two effects of addition of very small (.025%) amounts of anionic or cationic surfactant to a system stabilized by a nonionic surfactant. The expected effect on increase of the temperature at which maximum solubility was observed was found (the HLB-temperature). In addition a pronounced extension of the solubility area was revealed. It reached to considerably lower surfactant concentrations than in the case of the pure nonionic surfactant; it coalesced with the L_2 region and the W/O ratio range for its existence was grossly enlarged (Figs. 3,4,5).

On the other hand addition of an ampholytic surfactant at its isoelectric point or of an internal quaternary ammonium surfactant gave no influence on the solubility area at these low concentrations.

These results point to an influence of the ionic group; it will be separately evaluated for the capacity to reduce the minimum amount of surfactant needed and to enlarge the surfactant phase solubility region at the HLB-temperature.

A. Minimum surfactant concentration.

The evaluation of direct effects of the type of ionic group on the minimum amount of surfactants needed to form the microemulsion is facilitated by the use of molar concentrations. A collation of the necessary data is given below.

Those surfactant additives, dodecyl-β-alanine salt, and sodium dodecyl ether sulfate, which caused a rather small shift in HLB-temperature, also lead to only a slight increase in temperature stability in the range of the minimum surfactant concentration. On the other hand the sodium dodecyl sulfate, the sodium laurate and the dodecyl ammonium chloride show temperature ranges of the order of 5°C of the isotropic solutions at their lowest surfactant concentrations.

The sodium dodecyl ether sulfate was the most effective additive and the sodium dodecyl-β-alanine the least effective additive on a molar basis by regard to its ability to decrease the needed amount of surfactant to yield isotropic solutions containing equal amounts of oil and water. The results hence indicate that the ionic surfactant either should have its ionic group situated just at the interface of the hydrocarbon chains and the hydrophilic groups and if so be of as small size as possible or be situated at the interface of the hydrophilic groups and the water in order to stabilize the microemulsions optimally.

B. The solubility region of the surfactant phase.

The phase diagrams determined at 35°C for water-hexadecane-surfactant, where the surfactant consists of TEGDE and sodium dodecyl sulfate or dodecyl ammonium chloride in the weight ratio 199:1, indicated the presence of ionic surfactant to cause an extension of the surfactant phase to give a coalescence with the L_2-phase. The surfactant phase was also extended towards lower surfactant concentrations (Fig. 5) and to a wider range of W/O ratios.

A rational explanation of this behavior has to be based on the conditions in the surfactant phase and in micellar solutions. The following references are useful in this respect.

The stability of inverse (19-22) and normal micelles depend on the intermolecular Van der Waals potentials and on the forces between the ionic groups. Experimental illustrations to this influence is found both for liquid crystals and for micellar solutions.

Gulik-Krzywicki, et al, (23) have shown that the addition of ionic surfactants in the concentration range of 0.3-5% of lecithin lead to the lamellar phase formed with water being able to include water much above the maximum hydration of the pure lecithin. This effect became noticable at about 0.2% ionic surfactant added. If anionic and cationic surfactants are present in equivalent amounts no extra swelling was observed. This fact and also that addition

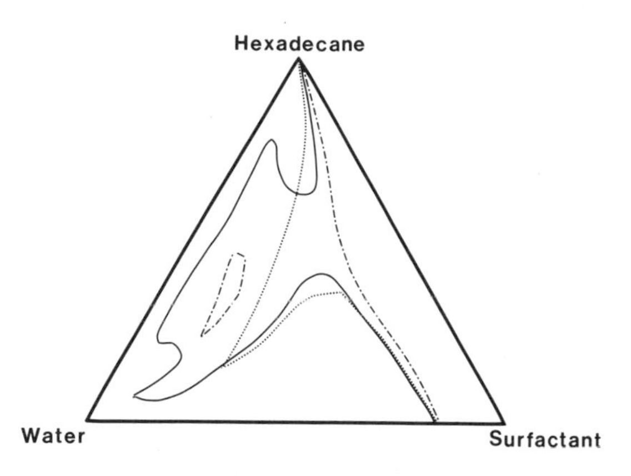

Fig. 5. Comparison of regions of isotropic liquids in systems of water, hexadecane and a surfactant.
Surfactant: tetraethylene glycol dodecyl ether 30°C (—.—.—); 35°C (.....).
Surfactant: tetraethylene glycol dodecyl ether and dodecyl ammonium chloride in weight ratio of 199:1. 35°C (———).

of simple electrolytes reduced the effect of the charged surfactant indicated that the forces from the compression of electric double layer was the stabilizing factor. Similar results have been reported by Krog and Borup (24) who also concluded that the increased swelling of the lamellar liquid crystal was due to an electric repulsion between the charged groups at the surface of the bilayer.

These results were concerned with the interlamellar distance in a liquid crystalline phase which means that the repulsion from the double layer influenced the maximal separation of layers with high transversal adhesion. The ion-ion repulsion forces within each layer were obviously not of a magnitude to cause destabilization of the lamellar structure to give normal micelles.

Our present results are concerned with the surfactant phase; the structure of which is so far not known with high reliability. Shinoda (25) has suggested a lamellar structure as a structure with properties between normal and inverse micelles. Such a structure cannot be in the form of a liquid crystalline phase since the surfactant is an isotropic liquid; an impossibility for the ordered lamellar arrangement of the liquid crystal. Furthermore, the surfactant phase is formed in equilibrium with the lamellar liquid crystal in nonionic systems (15). Scriven (27) has suggested a structure consisting entirely of saddle surfaces in order to minimize the first moment of the interfacial free energy and the present authors (15) an irregular structure containing both water and oil droplets in a kinetic equilibrium; the reason for stability being an extremely low interfacial tension giving the entropic contribution dominance.

Available experimental information (28,29) indicates the structure to consist of fairly large droplets of spherical or polyhedral shape. The important feature of the structure is its sensitivity even to minor intermolecular forces; a good example is the fact that a change of hydrocarbon from hexadecane to decane reduced the HLB-temperature by 5°C (13). The normal micelles observed at temperatures below the HLB-temperature do not show the corresponding properties. Schick and Manning (26) in their studies of micelle formation in mixtures of various polyethylene glycol dodecyl ethers and sodium dodecyl sulfate actually found that the degree of ionic repulsion of the ionic component markedly decreased as the proportion of the nonionic surfactant reached a threshold range of 10 mole %.

At 35°C, the pure TEGDE system formed W/O-microemulsions (Fig. 5) containing inverse micelles, meaning a tendency for the hydrophilic layer to be compressed compared to the hydrophobic hydrocarbon chain layer. Addition of small amounts of ionic surfactant gave a repulsion between the ionic surfactant molecules evidently sufficient to counteract this compression of the hydrophilic layer and a surfactant phase was formed. Such an interpretation postulates a structure in which the counter ions are dissociated from the surfactant molecules. The results by Meguro, et al., (30) who studied the packing of the ethylene glycol chains of heptaethylene glycol n-dodecyl ether (HEGDE) in mixed micelles with sodium dodecyl sulfate indicate that this is the case. From pNa and conductivity measurements they concluded that at molar ratios of $NaC_{12}SO_4$ to HEDGE larger than 1:1 the polyoxyethylene chains are loosely packed and a considerable ion binding takes place between the sodium ions and the dodecyl sulfate groups. When the relative concentration of HEGDE was increased in excess of this value the sodium ions were released indicating a shielding of the dodecyl sulfate groups from the sodium ions by the ethylene glycol chains.

At the molar ratios present in our systems we can therefore expect all sodium ions to be dissociated and an estimation of the repulsion forces from the ionic groups may be made using the molecular area for the sulfate group as 53.4 Å^2 (30). The molecular area for a densely packed polyethylene oxide chain with 4 ethylene oxide units can be estimated to be 50 Å^2 from Lange's data (31).

Assuming a hexagonal packing (32) the average distance between the $C_{12}SO_4$ ions will approximately be 4 nm when the molar ratio of TEGDE to $NaC_{12}SO_4$ is 120:1. If we assume the dielectric constant (ε) of the polyethylene oxide medium to be 30 (ε of glycol is 38) the repulsion energy between the electric charges can be estimated from

$$\Delta E_R = \int_{\infty}^{d_1} F dr = - \int_{\infty}^{d_1} \frac{e^2}{\varepsilon \cdot r^2} \cdot dr =$$

$$= \Big[_{\infty}^{d_1} \frac{e^2}{\varepsilon \cdot r} = \Big[_{\infty}^{4.10^{-7}} \frac{(3\cdot 10^9 \cdot 1.6\cdot 10^{-19})^2}{30 \cdot r} =$$

$$= \left[\frac{4.8^2 \cdot 10^{-20}}{30} \left(\frac{1}{4\cdot 10^{-7}} - \frac{1}{\infty}\right)\right] =$$

$$= \frac{4.8^2 \cdot 10^{-20}}{30 \cdot 4 \cdot 10^{-7}} \simeq 2\cdot 10^{-14} \text{ erg}$$

By dividing ΔE_R with kT (4.10^{-14} erg) we obtain a value of 0.5. A comparison with Table I, reference 10 shows that the repulsion energy is of the same order as or larger than the bending

Table I

Surfactant	Molar Ratio TEDGE: Surf.	Minimum amount of surfactant needed to yield Isotr. sol. W-%	"HLB-Temp" °C
$C_{12}H_{25}O-(CH_2CH_2O)_4-H$(TEGDE)	00	12	30
$C_{12}H_{25}OSO_3^-Na^+$	125:1	6	36
"	62:1	5.5	34,41
"	31:1	10.5	34,~46
$C_{11}H_{23}COO^-Na^+$	97:1	6	33
$C_{12}H_{25}O-(CH_2CH_2O)_3SO_3^-Na^+$	182:1	6	32
$C_{12}H_{25}NH_3^+Cl^-$	96:1	5	36
$C_{12}H_{25}NH_2CH_2CH_2COOH$	112:1	12	30
$C_{12}H_{25}\overset{+}{N}H_3CH_2CH_2COOH\ Cl^-$	128:1	6	32.5
$C_{12}H_{25}NH_2CH_2CH_2COO^-Na^+$	122:1	8.5	31
4 parts $C_{12}A$ + 1 part $C_{12}A^+Cl^-$	(334:1)*	9	31
4 parts $C_{12}A$ + 1 part $C_{12}A^-Na^+$	(318:1)*	12	30
$C_{12}H_{25}{}^+N(CH_3)_3CH_2CH_2COO^-$	118:1	12	30

*TEGDE: ionic surf.

energy (33) for those cases a considerable number of molecules can be located in a spherical micelle. Hence the repulsion from the surfactant ions appear sufficient to change the structure of a large inverse micelle with small bending energy; the small inverse micelles for which the bending energy component of the interfacial free energy is considerable (33) no significant change was observed (Fig. 3B).

The interesting results by Becher (34,35) on the high tentials of emulsion droplets stabilized by nonionic surfactants are of interest in this context. These results were obtained in an O/W emulsion; the present case is concerned with inverse micelles. No information on an internal potential in these systems is available.

CONCLUSIONS

The influence of small additions of an ionic surfactant to a nonionic system close to the HLB-temperature has been given a rational explanation as arising from repulsion between the ionic groups of the surfactant.

REFERENCES

1. G. Gillberg and S. Friberg, ACS Symposium on Evaporation-Combustion of Fuel Droplets, San Francisco, California, August (1976).
2. S. Friberg, Chemtech 6, 124 (1976).
3. L. M. Prince, J. Colloid Interface Sci., 52, 182 (1975).
4. A. W. Adamson, J. Colloid Interface Sci., 29, 261 (1969).
5. G. Gillberg, H. Lehtinen, and S. Friberg, J. Colloid Interface Sci., 33, 40 (1970).
6. K. Shinoda, and H. Kunieda, J. Colloid Interface Sci., 42, 381 (1973).
7. S. I. Ahmed, K. Shinoda, and S. Friberg, J. Colloid Interface Sci., 47, 32 (1974).
8. K. Shinoda and S. Friberg, Advances in Colloid and Interface Sci., 4, 281 (1975).
9. K. Shinoda and T. Ogawa, J. Colloid Interface Sci., 24, 56 (1967).
10. K. Shinoda, J. Colloid Interface Sci., 24, 4 (1967).

11. K. Shinoda and H. Saito, J. Colloid Interface Sci., 26, 70 (1968).
12. H. Saito and K. Shinoda, J. Colloid Interface Sci., 35, 359 (1971).
13. S. Friberg and I. Lapczynska, Progr. Colloid Polymer Sci., 56, 16 (1975).
14. I. Lapczynska and S. Friberg, 48th Nat. Colloid Symp., ACS, Austin (1974) Preprints, p. 168.
15. S. Friberg, I. Lapczynska and G. Gillberg, J. Colloid Interface Sci., 56, 19 (1976).
16. J. B. Brown, I. Lapczynska and S. Friberg, Proceedings of the International Conference on Colloid and Surface Science, Budapest, Sept. 1975, Preprints, p. 507, ed., E. Wolfram.
17. L. Mandell and P. Ekwall, Acta Polytechn. Scand. Ch. 741 (1968).
18. T. L. Gresham, et al., J. Amer. Chem. Soc., 73, 3168 (1951).
19. H. F. Eicke and H. Christen, J. Colloid Interface Sci., 46, 417 (1974).
20. Idem, Ibid., 48, 281 (1974).
21. E. Ruckenstein and J. C. Chi, J. Chem. Soc. Faraday Trans. II, 71, 1690 (1975).
22. S. Levine and K. Robertson, J. Phys. Chem., 76, 876 (1972).
23. T. Gulik-Krzywicki, A. Tardieu and V. Luzzati, Mol. Cryst. Liq. Cryst., 8, 285 (1969).
24. N. Krog and A. P. Borup, J. Sci. Fd. Agric., 24, 691 (1973).
25. H. Saito and K. Shinoda, J. Colloid Interface Sci., 32, 647 (1970).
26. M. J. Schick and D. J. Manning, J. Amer. Oil Chem. Soc., 43, 133 (1966).
27. C. A. Miller and L. E. Scriven, J. Colloid Interface Sci., 33, 360 (1970).
28. D. O. Shah--Personal Communication.
29. K. Madani--Personal Communication.
30. K. Meguro, H. Akasu and M. Ueno, J. Amer. Oil Chem. Soc., 53, 145 (1976).

31. H. Lange, Kolloid-Z. Z. Polymere, 201, 131 (1965).
32. E. Murphy, Ph.D. Thesis, University of Minnesota, 1966.
33. D. N. Sutherland, J. Colloid Interface Sci., 60, 96 (1977).
34. P. Becher and S. Tahara, Proc. VIth Int. Congr. Surface Activity, Zurich, II2, 519 (1972).
35. P. Becher, S.E. Trifiletti, and Y. Machida in "Theory and Practice of Emulsion Technology, SCl Symp., September, 1974, Brunel Univ.," A.L. Smith (ed.), Academic, London, 1976, pp. 271-280

DISCUSSION

Sydney Ross (Rensselaer Polytechnic Institute)

In a three-component system at a fixed temperature, three phases can co-exist only if the composition of each of them is fixed. That region of compositions in which the three phases are present has to be represented in a triangular phase diagram as an internal triangle, in which each vertex represents, respectively a water-rich phase, an oil-rich phase, and a surfactant-rich phase. The surfactant-rich phase is represented by a point on the periphery of a one-phase island. If that island should denote a single phase that happens to be a liquid-crystal phase, then stable emulsions are formed when the three phases co-exist, according to your findings. Am I correct then in supposing in such a case that only within the range of compositions represented by points within the internal triangle, can be found stable emulsions of the type you are describing?

Author The assumption is correct. When the number of phases changes from 2 liquid phases to 2 liquid phases and 1 liquid crystalline phase a pronounced increased stability is observed.

It is essential to realize that many two-phase emulsions are very stable indeed and that the stability due to the presence of a liquid crystalline phase is a special case.

A. S. Kertes (Hebrew University, Jerusalem, Israel)

Is there any effect of temperature on the double spacing in your x-ray data of the third phase? If yes, in what direction?

Would you have some numerical data to illustrate this effect or the lack of it?

Author We have not determined the temperature dependence of the x-ray data. The results would depend on the composition changes of the third phase with temperature. For a system water-oil-lecithin the changes would be small; for a nonionic system, on the other hand, a pronounced decrease of the spacing would certainly be experienced in some systems in certain temperature ranges.

A. Silberberg (Weizmann Institute)

In what way is your third phase different from interfacial phases in general? Near to critical conditions interfacial phases tend to become infinitely thick. Moreover it is easily possible to obtain metastable states.

Author The difference is that the third phase reported in this paper has an inherent stability independent of the presence of a stabilizing interface i.e., the phase can be separated from the system and its structure determined.

The comparison with critical conditions is interesting in connection with the isotropic liquid phase; its structure may be similar to a liquid close to the critical condition.

DIELECTRIC BEHAVIOUR OF WATER-IN-UNDECANE MICROEMULSIONS USING NONIONIC SURFACTANTS

J. Peyrelasse, C. Boned, P. Xans and M. Clausse

Laboratoire de Thermodynamique
Institut Universitaire de Recherche Scientifique
Université de Pau et des Pays de l'Adour
Pau, France

A B S T R A C T

The dielectric behaviour of water-in-undecane microemulsions, using a blend of octylphenylether polyoxyethylenes as the surfactant, was investigated versus their temperature T and mass fraction of water p.

It was found that these systems exhibit, along with a conduction absorption, a Cole-Cole type dielectric relaxation whose features are dependent on both T and p. When increasing T, at a given value of p, the low frequency conductivity of the samples decreases, then increases, its minimum value corresponding to the solubilization end temperature.

An explanation of these phenomena is proposed that relies on possible structural changes occurring in microemulsions when their temperatures are raised.

I. INTRODUCTION

The mutual solubilization of water and hydrocarbons or halogenocarbons can be achieved by using non-ionic surfactants such as polyoxyethylene alkylphenols or alcohols. For instance, SHINODA and co-workers, [1, 2, 3, 4, 5], have studied ternary systems involving water, "oil" type liquids such as hexane, cyclohexane, isooctane, toluene, m-xylene or carbon tetrachloride and nonylphenylether polyoxyethylenes of different chain lengths. They investigated the dependence of the phase behaviour of these systems upon composition, temperature and surfactant concentration and chain length.

They showed in particular that, in the lower temperature range, transparent oil-swollen micellar systems result from the solubilization of a hydrocarbon or halogenocarbon in aqueous solutions of non-ionic surfactants. On the contrary, in the higher temperature range, water dissolves in non-aqueous solutions of non-ionic surfactants so as to form transparent water-swollen micellar systems. Although their obtention is submitted to very definite conditions of temperature and composition, these transparent micellar systems are very akin to the so-called microemulsions made of water, a hydrocarbon, an alcohol and a soap, [6].

Since they were first introduced by HOAR and SCHULMAN, [7], much attention has been paid to microemulsions because of their possible applications in numerous fields of technology and industry, [8, 9]. Their structure and properties were investigated by SCHULMAN and co-workers who used various techniques such as X-ray diffraction, light scattering, nuclear magnetic resonance, electron microscopy, ultracentrifugation, rheology and conductivity measurements, [10, 11, 12, 13, 14]. They concluded from their studies that microemulsions consist of dispersions of minute droplets less than 1 000 Å in diameter encased in a "shell" of combined surfactant and cosurfactant and that their formation occurs spontaneously owing to a metastable negative interfacial tension. These conclusions have been endorsed, for some variations, by other authors, [15, 16, 17, 18, 19, 20]. On this basis, SHAH and co-workers, [21, 22, 23], proposed a mechanism of phase inversion occuring in water and hexadecane systems using potassium oleate as the surfactant and hexanol as the cosurfactant.

A different approach has been followed by FRIBERG and others, [6, 8, 24], who assert that there is no need to distinguish microemulsions from micellar solutions, as some authors tried to, [25, 26], especially in the case of water-in-oil type systems for which they propose a model of phase behaviour, [27]. This opinion was supported by EKWALL, MANDELL and FONTELL, [28], who stressed the similarity between microemulsions and isotropic micellar solutions and concluded that the term "microemulsion" should not be used in connection with transparent oil-in-water or water-in-oil systems.

Studies of the conductivity and complex permittivity of different kinds of microemulsion systems, [29, 30, 31, 32], have shown that their dielectric properties do not fit in with the theoretical model of dielectric behaviour that has been established by different authors, [33, 34, 35, 36, 37], and has proved suitable for many emulsion systems, [33, 35, 37, 38, 39, 40, 41, 42, 43].

For instance, water-in-hexadecane transparent systems, the surfactant being potassium oleate and the cosurfactant hexanol, exhibit in the 10 MHz region a dielectric relaxation as ordinary water-in-oil emulsions do, but the values of the dielectric increment are by far too large compared with those that can be derived from the theoretical model. Moreover, it was found that their low frequency conductivity is fairly high, contrary to the case of ordinary water-in-oil emulsions.

In an attempt to give a further contribution to the understanding of the nature and properties of the so-called microemulsions, a dielectric study has been undertaken, which concerns water-in-oil transparent systems using non-ionic surfactants.

II. MATERIALS AND METHODS

A. Preparation of Samples

Samples were made from resin-exchanged water and undecane ("Purum" grade, from Fluka A. G.), the surfactant being a blend of two octylphenylether polyoxyethylenes of different chain lengths, (Octarox 1 and Octarox 5, from Montanoir, France).

The blend of Octarox 1 (10% wt/wt) and Octarox 5 (90% wt/wt) was mixed, at room temperature, with undecane so as to obtain an "oil-plus-surfactant" phase containing 12% (wt/wt) surfactant. Samples of various weight fractions p of water were prepared by adding various amounts of water to portions of this "oil-plus-surfactant" phase.

Following a method that has been described and used by SHINODA and others, [2, 3, 4, 5], the phase behaviour of the samples was studied under conditions of slow heating or cooling, in order to determine the realm of existence of transparent water-in-undecane systems and its boundaries, namely the cloud curve and the solubilization end curve.

B. Dielectric Measurements

Up to 3 MHz, the relative complex permittivity $\varepsilon^* = \varepsilon' - j\varepsilon''$ of the samples was determined, with an uncertainty of ± 0.01 for both ε' and ε'', by means of an apparatus consisting essentially of two Schering type admittance bridges (General Radio 716 C and 716 CS 1) used in conjunction with either General Radio 1210 C or 1211 C oscillators, depending on the frequency range investigated, and with a General Radio 1232 A detector unit, the test cell being a thermostatically controlled cylindro-conical capacitor especially designed for liquids (Ferisol CS 601).

Determinations of ε' were also made at 2100 MHz by means of a Narda 231 N coaxial line used in conjunction with a Ferisol OS 301 A oscillator connected to a Ferisol SCF 201 power supply unit.

The low frequency conductivity χ_ℓ of the samples was studied versus temperature, using Mullard cells, (Philips), connected either to a General Radio 1680 automatically balancing admittance bridge working at 400 Hz and 1 KHz or to a Wayne-Kerr precision conductance bridge working at an angular frequency of 10^4 rad/s.

The variations of the temperature T were checked by means of a calibrated thermocouple, with an uncertainty of ± 0.2°C.

III. EXPERIMENTAL RESULTS

As it has been reported in the preceding section, the realm of existence of transparent water-in-undecane systems was established by investigating the phase behaviour of the samples under conditions of slow heating or cooling. For different values of the mass fraction p of water, the temperatures corresponding to the cloud point and to the solubilization end point were determined. It was thus possible to obtain both the cloud curve C and the solubilization end curve S, as it is shown on Fig. 1.

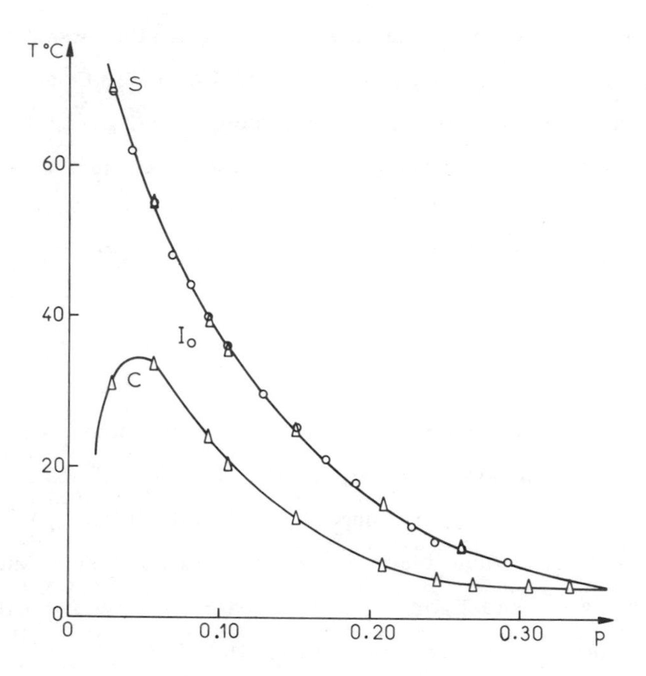

Fig.1 . The solubilization of water in undecane containing 12% non-ionic surfactant. C is the cloud curve and S the solubilization end curve. I_o is the realm of existence of stable transparent water-in-undecane systems. (The circles indicate the determinations of solubilization end points obtained from conductivity measurements).

The general features of the diagram are consistent with those reported for similar systems by different authors, [2, 3, 4, 5, 44]. I_o represents the realm of existence of water micellar solutions in undecane. These systems, which are transparent and stable when maintained at an adequate temperature, will be referred to as water-in-undecane microemulsions. When the temperature of such a microemulsion is raised above its solubilization end point, turbidity occurs which is followed by a separation process that leads to the splitting of the system into

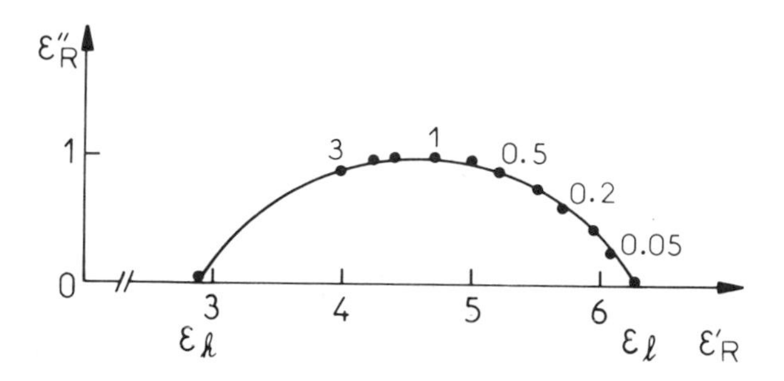

Fig. 2 . Cole-Cole plot of a water-in-undecane microemulsion. p = 0.093 , T = 36°C , $\chi = 7.9\ 10^{-6}\Omega^{-1}m^{-1}$. *(The frequency* ν *is given in MHz).*

two phases. When the temperature of a microemulsion is lowered below its cloud point, the system becomes hazy and eventually splits into two distinct phases.

Experiments showed that, all over I_0, water-in-undecane microemulsions exhibit, along with a conduction absorption, a dielectric relaxation, [45]. Fig. 2 gives an example of a Cole-Cole plot corresponding to $p = 0.093$ and $T = 36\,°C$.

At each frequency, the dielectric relaxation loss factor ε''_R was obtained by subtracting from the global loss factor ε'' the conduction loss factor ε''_C derived from the determination of the low frequency conductivity χ_ℓ. Owing to the weakness of χ_ℓ, no special correction was applied to ε'_R whose values were assimilated to those of ε'. As for ε_h, its value, measured at 2100 MHz by means of a coaxial line, is correctly located on the Cole-Cole plot. By using SMYTH's method, [46], it was checked that the distribution of the experimental points on the Cole-Cole plot fits in with a dielectric relaxation of the Cole-Cole type.

Consequently, the complex relative permittivity ε^* of water-in-undecane microemulsions can be written :

$$\varepsilon^* = \varepsilon_h + \frac{\varepsilon_\ell - \varepsilon_h}{1 + j(\omega/\omega_c)^{(1-\alpha)}} + \frac{\chi_\ell}{j\omega\varepsilon_o} \quad (1)$$

j being the square root of the negative unit, ε_o the permittivity of free space and ω the angular frequency, related by $\omega = 2\pi\nu$ to the frequency ν of the applied electrical field.

χ_ℓ , the low frequency conductivity, ε_h , the high frequency limiting relative permittivity , ε_ℓ , the low frequency limiting relative permittivity, $\nu_c = \omega_c / 2\pi$, the critical frequency, and α , the Cole-Cole coefficient, depend on both T and p .

Fig. 3 shows as an example the variations of χ_ℓ versus T observed for a microemulsion whose mass fraction of water is $p = 0.056$.

As T increases , the conductivity χ_ℓ of a water-in-undecane microemulsion decreases down to a minimum and then increases. The temperature T_m , at which χ_ℓ reaches its minimum value χ_m was found to be equal to the solubilization end temperature, whichever the value of p . Thus it has been possible to redetermine with a good precision the solubilization end curve, as it is shown on Fig. 1 .

As concerns the dielectric relaxation, experiments showed that there is no meaningful variation of ε_h when the temperature varies while ε_ℓ and, consequently, the dielectric increment $(\varepsilon_\ell - \varepsilon_h)$ are decreasing functions of T . The critical frequency ν_c is an increasing function of T , α remaining nearly constant. To illustrate this behaviour, results are reported in the table thereafter which is relevant to a microemulsion characterized by $p = 0.056$.

The influence of p upon the dielectric properties of water-in-undecane microemulsions is rather complex because of its connection with the phase diagram. On Fig. 4 , the variations of χ_ℓ versus T are plotted for different values of the mass fraction p of water.

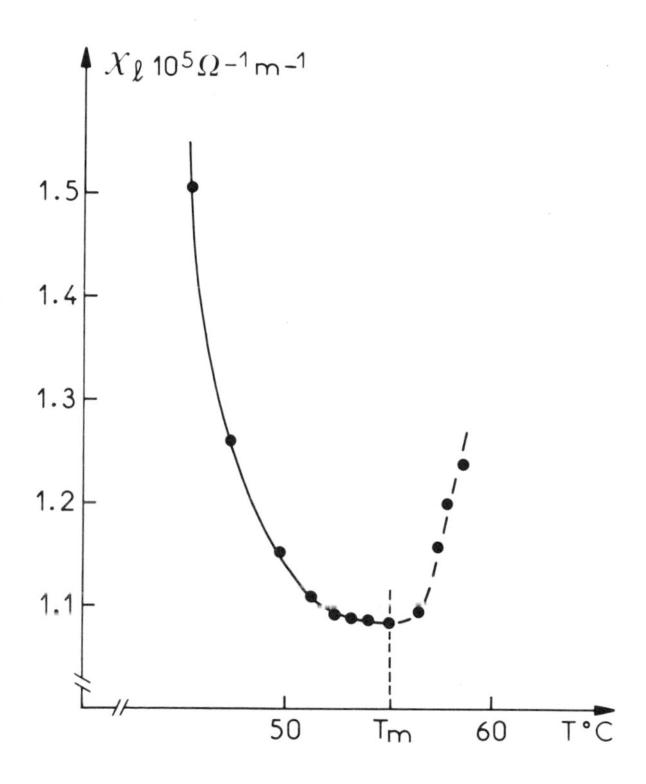

Fig. 3 . Variations of χ_ℓ versus T for p = 0.056

T A B L E

Influence of T on the Dielectric Relaxation, for p = 0.056

T (°C)	$10^4 \chi_\ell$ ($\Omega^{-1} m^{-1}$)	ε_ℓ	ε_h	$\varepsilon_\ell - \varepsilon_h$	ν_c (MHz)	α
36.3	1.10	9.20	2.40	6.80	0.45	0.39
38.2	0.71	8.80	2.55	6.25	0.47	0.38
40.1	0.45	8.00	2.60	5.40	0.54	0.38
41.6	0.31	7.10	2.55	4.55	0.63	0.41

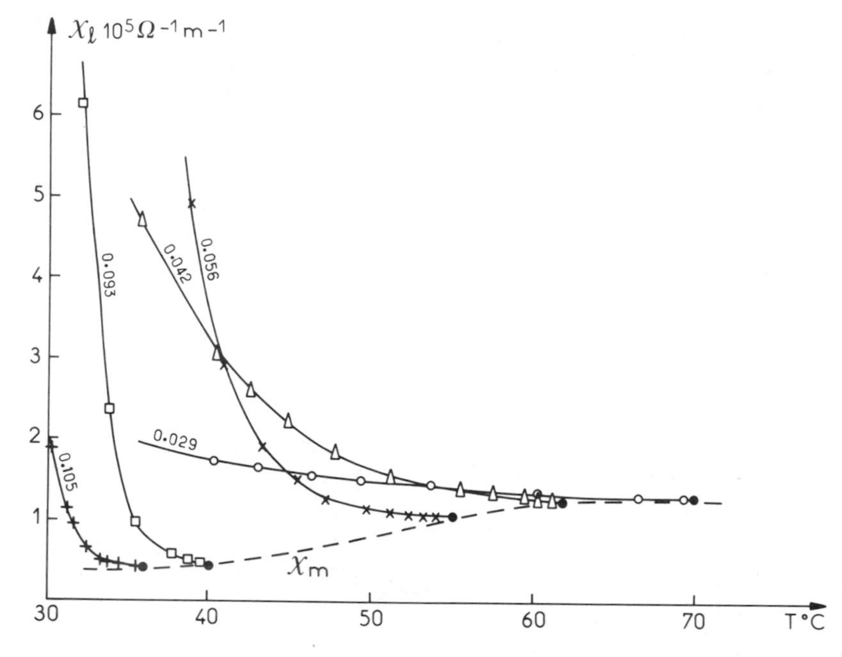

Fig.4 . Variations of χ_ℓ *versus T for different values of p*

Upon decreasing p , the curves representing the variations of χ_ℓ versus T become flatter while χ_m increases and shifts towards higher temperatures.

IV. DISCUSSION

As reported by several authors, [33, 35, 38] , water-in-oil type emulsions involving non-polar and non-conductive oil type phases are themselves non-conductive but exhibit a Cole-Cole dielectric relaxation related to migration polarization phenomena. This behaviour has been satisfactorily explained from the "spherical dispersion" model and the formula proposed by HANAI, [33], to describe it,

$$\left(\frac{\varepsilon^*_1 - \varepsilon^*_2}{\varepsilon^*_1 - \varepsilon^*}\right)^3 \frac{\varepsilon^*}{\varepsilon^*_2} = \frac{1}{(1-\Phi)^3} \qquad (2)$$

where ε^*_1 , ε^*_2 and ε^* represent respectively the relative complex permittivities of the disperse phase, the continuous phase and the emulsion, Φ being the volume fraction of the disperse phase. It has been shown by CLAUSSE and ROYER, [35, 37], from computer calculations that, in the case of emulsions made of an aqueous phase dispersed within a non-polar and non-conductive continuous oil phase, Eq. (2) can be written like a Cole-Cole formula :

$$\varepsilon^* = \varepsilon_h + \frac{\varepsilon_\ell - \varepsilon_h}{1 + (j\,\omega/\omega_c)^{(1-\alpha)}} \qquad (3)$$

ε_h, ε_ℓ and $\varepsilon_\ell - \varepsilon_h$ are increasing functions of Φ , but their variations upon the temperature T are negligible. ω_c is a decreasing function of Φ and an increasing function of T with an activation energy equal to the activation energy of conductivity of the disperse phase. Whichever the values of Φ and T , α is small compared with 1 .

From the results reported in the preceding section, it appears that the dielectric behaviour of water-in-undecane microemulsions differs from that of ordinary water-in-oil emulsions. Concerning the features of the dielectric relaxation, it should be noted in particular that the Cole-Cole plots of water-in-undecane microemulsions are more depressed than those observed for water-in-oil emulsions, the coefficient α being equal more or less to 0.40 while it reaches a maximum of 0.06 only in the case of water-in-oil emulsions, [35]. Moreover, it can be seen from the table that ε_ℓ and $(\varepsilon_\ell - \varepsilon_h)$ are strongly dependent upon T , contrary to the case of emulsions. A similar dependency of the dielectric increment $(\varepsilon_\ell - \varepsilon_h)$ upon the temperature has been found by EICKE and SHEPHERD, [47], who investigated the dielectric properties of solubilized water in solutions of sodium-di-2-pentyl-sulfosuccinate, (Aerosol AY), in benzene. These authors observed also that the critical frequency of the dielectric relaxation is an increasing function of T.

An additional support to this opinion can be drawn from the low frequency conductivity measurements made on water-in-undecane microemulsions. The mere fact that these stable transparent systems are conductive indicates that part of the water is not engaged in aggregates, (either micelles or droplets, depending on the view-point), and is solubilized at a molecular level in the continuous phase. In the case of water-in-toluene microemulsions using sodium dodecyl sulphate as the surfactant and butanol as the cosurfactant, some authors, [48], have claimed that the continuous phase contains about 0.3 % water.

The decrease of χ_ℓ versus T could suggest that, as the temperature increases, a repetitive aggregational process takes place within the system, individual molecules of water being driven to micellar aggregates that gather to form larger micelles and eventually droplets, when T approaches the solubilization end temperature T_m. At T_m, the microemulsion would be the most "emulsion-like", a further small increase of T leading to its splitting into two phases and, consequently, to an increase of the conductivity of the system. This tentative explanation of the conductivity phenomena observed for water-in-undecane transparent systems is to be considered in connection with the results obtained by EICKE and REHAK, [49], who followed, by light scattering and ultracentrifugation measurements, the formation of water-in-isooctane microemulsions using sodium-di-2-ethyl-hexyl-sulfosuccinate, (Aerosol AOT). They concluded from their experiments that, with increasing water concentration, the microemulsion is stabilized by repeated aggregational processes of water-swollen micelles due to a decrease of the interfacial free enthalpy.

In a recent paper, [50] , EICKE pointed out that consideration should be given to the dynamic aspect of microemulsion properties, in particular to exchange processes of solubilized material between micelles. It is worth mentioning also that SKOULIOS and GUILLON, [51], proposed to consider the stability of microemulsions to be eminently dynamic because of the competition of micellar order and osmotic order within them.

A C K N O W L E D G M E N T S

The authors wish to express their thanks to Mrs. ANGLICHAU, Mrs. M. BERNADE and Mr. C. CERESUELA who prepared the typescript and the illustrations of this paper.

V. R E F E R E N C E S

1 . K. Shinoda, in Solvent properties of surfactant solutions K. Shinoda, ed.), Chap. 2 , pp. 27-63 . Dekker, New-York , 1967 .

2 . K. Shinoda and T. Ogawa, J. Colloid Interface Sci., 24 : 56 (1967) .

3 . K. Shinoda and H. Saito, J. Colloid Interface Sci., 26 : 70 (1968) .

4 . H. Saito and K. Shinoda, J. Colloid Interface Sci., 32 : 617 (1970) .

5 . H. Saito and H. Kunieda, J. Colloid Interface Sci., 42 : 381 (1973) .

6 . K. Shinoda and S. Friberg, Advances Colloid Interface Sci. , 4 : 281 (1975) .

7 . T. P. Hoar and J. H. Schulman, Nature (London) , 152 : 102 (1943) .

8 . S. Friberg, Informations Chimie, 148 : 235 (1975) .

9 . L. M. Prince, in Emulsions and Emulsion Technology, (K.J. Lissant, ed.), Part I , Chap. 3 , pp. 125-177 . Dekker, New-York (1974) .

10 . J. H. Schulman and D. P. Riley, J. Colloid Sci., 3 : 383 (1948) .

11 . J. H. Schulman and J. A. Friend, J. Colloid Sci., 4 : 497 (1949) .

12 . J. E. L. Bowcott and J. H. Schulman, Z. Elektrochem., 59 : 283 (1955)

13 . W. Stockenius, J. H. Schulman and L. M. Prince, Kolloid-Z., 169 : 170 (1960) .

14 . C. E. Cooke and J. H. Schulman, in Proc. 2nd Scandinavian Symp. Surface Activity, p. 231 , Stockholm , 1965 .

15 . L. I. Osipow , J. Soc. Cosmetics Chemists, 14 : 277 (1963) .

16 . L. M. Prince, J. Colloid Interface Sci., 23 : 165 (1967) .

17 . L. M. Prince, J. Colloid Interface Sci., 2 : 216 (1966) .

18 . L. M. Prince, J. Soc. Cosmetic Chemists, 21 : 193 (1970) .

19 . W. Gerbacia and H. L. Rosano, J. Colloid Interface Sci., 44 : 242 (1973) .

20 . H. L. Rosano, J. Soc. Cosmetic Chemists , 25 : 609 (1974) .

21 . D. O. Shah and R. M. Hamlin Jr. , Science , 171 : 483 (1971) .

22 . D. O. Shah, A. Tamjeedi, J. W. Falco and R. D. Walker Jr. AIChE Journal , 18 : 1116 (1972) .

23 . J. W. Falco, R. D. Walker and D. O. Shah , AIChE Journal, 20 : 510 (1974) .

24 . G. Gillberg, H. Lehtinen and S. Friberg, J. Colloid Interface Sci. , 33 : 40 (1970) .

25 . L. M. Prince, J. Colloid Interface Sci. , 52 : 182 (1975) .

26 . D. O. Shah, R. D. Walker, W. C. Hsieh, N. J. Shah, S. Dwivedi, J. Nelander, R. Pepinski and D. W. Deamer, Preprint SPE 5815 , Improved Oil Recovery Symposium of the Society of Petroleum Engineers of AIME , Tulsa , Okla. , U. S. A. , 1976 .

27 . S. Friberg, Report to D. G. R. S. T. , Paris , 1976 .

28 . P. Ekwall, L. Mandell and K. Fontell, J. Colloid Interface Sci. , 33 : 215 (1970) .

29 . M. Clausse and P. Sherman, C. R. Acad. Sci. , Ser. C 279 : 919 (1974) .

30 . M. Clausse, P. Sherman and R. J. Sheppard, J. Colloid Interface Sci., 56 : 123 (1976) .

31 . M. Clausse, R. J. Sheppard, C. Boned and C. G. Essex, in Colloid and Interface Science (M. Kerker, ed.), Vol. II, pp. 233-243, Academic Press, New-York, 1976 .

32 . R. A. Mackay, C. Hermansky and R. Agarwal, in Colloid and Interface Science (M. Kerker, ed.), Vol. II, pp. 289-303, Academic Press, New-York, 1976 .

33 . T. Hanai, in Emulsion Science (P. Sherman, ed.), Chap. 5, pp. 353-478, Academic Press, London, 1968 .

34 . S. S. Dukhin, in Surface and Colloid Science (E. Matijevic, ed.), Vol. III, pp. 83-165, Wiley Interscience, New-York, 1971 .

35 . M. Clausse, Thesis, University of Pau, France, 1971 .

36 . M. Clausse, C.R. Acad. Sci., Ser. B 274 : 649 (1972) .

37 . M. Clausse and R. Royer, in Colloid and Interface Science (M. Kerker, ed.), Vol. II, pp. 217-232, Academic Press, New-York, 1976 .

38 . C. Lafargue, M. Clausse and J. Lachaise, C.R. Acad. Sci., Ser. B 274 : 540 (1972) .

39 . M. Clausse, C.R. Acad. Sci., Ser. B 274 : 887 (1972) .

40 . M. Clausse, C.R. Acad. Sci., Ser. B 277 : 261 (1973) .

41 . M. Clausse, Colloid Polymer Sci., 255 : 40 (1976) .

42 . M. Clausse and J. Lachaise, J. Phys. D : Appli. Phys., 8 : 1227 (1975) .

43 . C. Lafargue, J. Lachaise and M. Clausse, in Proc. 8th International Conference on Properties of Water and Steam, pp. 492-504, 1975 .

44 . M. L. Robbins, in Symposium on Advances in Petroleum Recovery, pp. 297-325, New-York, 1976 .

45 . J. Peyrelasse, C. Boned, P. Xans and M. Clausse, C. R. Acad. Sci., Ser. B 284 : 235 (1977) .

46 . C. P. Smyth, in Dielectric Behaviour and Structure, Mc Graw Hill, New-York , 1955 .

47 . H. F. Eicke and C. W. Shepherd, Helvetica Chimica Acta , 57 : 1951 (1974) .

48 . A. Graciaa, J. Lachaise, A. Martinez, M. Bourrel and C. Chambu, C. R. Acad. Sci. , Ser. B 282 : 547 (1976) .

49 . H. F. Eicke and J. Rehak, Helvetica Chimica Acta, 59 : 2883 (1976) .

50 . H. F. Eicke, in Colloid and Interface Sci. (M. Kerker, ed.) , Vol. II , p. 319 ; Academic Press , New-York , 1976 .

51 . A. Skoulios and D. Guillon, J. Physique Lettres, 38 : L-137 (1977).

EXPERIMENTAL RESEARCHES ON SILICONE ANTIFOAMS

Sydney Ross and Gary Nishioka

Department of Chemistry
Rensselaer Polytechnic Institute, Troy, N. Y.

INTRODUCTION

Silicone oil (polydimethylsiloxane) is an effective antifoam for oil foams, and was used during the Second World War to prevent the foaming of lubricating oil in aircraft engines. (1) But it is far from being equally as effective to control foaming in aqueous solutions. It was found, however, that after the oil is "thickened" by the admixture of finely divided silica in proportions of 3 to 6%, it acts as a highly effective antifoam, even when only a few parts per million of the mixture is present in the foamy liquid. The scientific basis for the promotion or "activation" of the silicone oil by the presence of the silica has never received adequate study and many misunderstandings about it are current. This unanswered question has not hindered the production of scores of patents describing progressively better methods to incorporate the silica into the formulation, with more effective antifoaming action marking each advance in the art of preparation. The accumulated evidence reported in those published patents leaves no doubt of the reality of the "activation," though it contributes nothing to our understanding of why it should occur. The improvement of the antifoam by the presence of the silica is, however, now so well established that few, if any, commercial formulations designed to suppress or destroy the foam of aqueous solutions by means of silicone oil are marketed without the added silica.

Characteristics of Hydrophilic and Hydrophobic Silicas

Silica normally has a polar surface, which is to say it is perfectly wetted by water and separates readily into its ultimate particles without clumping or clotting when shaken up with water. But the surface of silica can be drastically altered by chemical treatment to become hydrophobic. Since fine-divided particles necessarily have a large surface area, and we are talking of silicas with 200 to 400 square meters of surface per gram, this alteration of the surface implies marked alterations of the gross behavior of the material. A particle of hydrophobic silica is not wetted by water; it floats on the surface of the water, just as would an oiled needle, in spite of its densiy being greater than that of water. Water may be said to roll off its back, just as it rolls off a duck's back, and fundamentally for the same reason. When mixed with a hydrophobic liquid, such as silicone oil, however, hydrophobic silica is readily wetted and disperses easily. In such a liquid medium it is now the turn of polar silica to show reluctance to be wetted; it does not allow the oil to penetrate freely along the surface and especially into the narrow spaces between particles. If many such polar particles are stirred into silicone oil, they may create an interconnected or network structure throughout the oil and so give it the viscoelastic properties of a jelly.

Dispersions of silicas in polydimethylsiloxane can be of two types, which we designate alpha (α) and beta (β), depending on how much heat or how much agitation is put into the system during its preparation. The α-dispersion is very viscous, and elastic, and resembles a jelly. We have prepared it, for example, from Aerosil 200 (Degussa Inc.), which was first dried at 150°C for a few hours, and then stirred into polydimethylsiloxane (1000 cstk. fluid) at room temperature at 3% solids. If the α-dispersion is ball-milled or heated to 150°C for a few hours, or to higher temperatures for a shorter time, the jelly-like structure breaks down irreversibly and the dispersion becomes more fluid. The dispersion is evidently different both to the eye and to the touch. This is the β-dispersion. The actual mode and details of preparation of the β-dispersion, that is, the kind of mixing or milling equipment, the ultimate temperature to which the dispersion was heated, or the periods of time given to these operations, are not uniquely combined as determining factors, because they can be varied widely in combinations that still attain the same end product. Thus, increased efficiency of milling can replace heating to a higher temperature, and prolonged heating at a lower temperature is equivalent in its effect to a shorter time of heating at a higher temperature.

The α and β dispersions differ in the following respects:
1) The obvious and immediately-apparent difference is the jelly-like character of the α-dispersion compared with the fluid, though still viscous, β-dispersion. If, for example the concentration of silica is made more than about 5% the α-dispersion is too stiff to be stirred conveniently by hand or by ordinary laboratory mixing equipment, but on conversion to the β-dispersion it flows like a heavy lubricating oil.
2) On diluting the α-dispersion with hexane, the dispersion of

silica becomes unstable and the solid soon separates and sinks to the bottom. This behavior reveals that the dispersion lacks true stability and is maintained only by virtue of its own supporting structure. The β-dispersion on the other hand is evidently well stabilized because on dilution with hexane to a low viscosity the particles of silica remain suspended indefinitely.

3) Tested as foam-inhibiting agents, the α-dispersion hardly differs from polydimethylsiloxane taken by itself, that is, without added silica; whereas the β-dispersion enhances foam destabilization, an effect well-recognized in the commercial compositions of polydimethylsiloxane as being conferred by silica.

4) The determination of "bound" polydimethylsiloxane by means of the film balance, (2) shows that the α-dispersion fixes much less polydimethylsiloxane to the silica surface than is fixed in the β-dispersion. Comparison of results gives for the α-dispersion about 0.10 g bound polydimethylsiloxane per gram of silica, and for the β-dispersion about 1.0 g bound polydimethylsiloxane per gram of silica.

5) When spread as a compressed monomolecular layer on a water surface, the α-dispersion stabilizes single bubbles blown under it exactly as does the monomolecular layer spread from polydimethylsiloxane taken by itself, that is, without added silica. A monomolecular layer spread from the β-dispersion does not stabilize a bubble at any degree of its compression.

These differences demonstrate that the silica in the α-dispersion is only partly converted to hydrophobic silica, and in that partially-converted form it adds nothing at all to the foam-inhibiting or fatigue-preventing properties of the polydimethylsiloxane.

The following commercial preparations of the polydimethylsiloxane containing silica, all of them marketed as foam inhibitors, were tested by the preceding five criteria. Each one was found to correspond to the β-dispersion by these tests.

Dow Corning Antifoam M
Dow Corning Antifoam MSA
Rhodorsil Antimousse 454

APPARATUS AND PROCEDURE

(1) Bubbles were created underneath a monolayer on a water substrate, contained in a Teflon-coated glass trough set in a dust-free box. The water used was doubly distilled from an all-Vycor apparatus. Before the polydimethylsiloxane was added, the water was swept several times and did not stabilize bubbles.

Table I

Properties of Silicas Used

Silica	Manufacturer	Measured Surface Area (m^2/g)	Manufacturer's Data		
			Surface Area (m^2/g)	Pore Volume (cm^3/g)	Average Aggregate Diameter (μm)
Cab-o-Sil M-5	Cabot Corp.	208	200	0	0.1 to 1.0
Aerosil 200	Degussa Inc.	237	200	0	0.1 to 1.0
Gasil 23	Crosfield Chem.	373	220	1.5	3 (porous)
Gasil 35	Crosfield Chem.	432	250	1.1	3 (porous)
Gasil 200	Crosfield Chem.	780	800	0.4	5 (porous)
Recovered from DC-MSA		207			
Recovered from DC-MSM		207			

Bubbles were delivered by means of a micrometer-driven syringe fitted to a Teflon-coated glass tube that had an inner diameter of 1 mm. They were formed approximately 10 mm. below the water surface and had an average volume of 2.46 cm.^{3} ± 3%. On reaching the surface their duration was timed with a stopwatch. Every value reported is a mean of twenty determinations.

(2) Dynamic foam stabilities were measured in an apparatus already described. (3) Foam inhibitor was injected into the foaming liquid through a rubber septum placed in the column wall.

(3) The Ross-Miles test was conducted according to ASTM recommendations. (4) The temperature of measurement was 20°C. instead of the recommended temperature of 120°F.

SOURCES OF MATERIALS

The sources and properties of materials used and the dispersions prepared are presented in Tables I, II, and III.

Table II

Number Average Molecular Weights of Polydimethylsiloxane

Polydimethylsiloxane	M_n From intrinsic viscosity	M_n From bulk viscosity	Manufacturer's data
Centrifugate from DC-MSA	2.6×10^4	2.5×10^4	--
DC-M	2.5×10^4	2.5×10^4	2.4×10^4
DC-200 (1000 cSt fluid)	2.5×10^4	2.5×10^4	--

Table III

Bound Polydimethylsiloxane on Silica Substrates[a]

Polydimethylsiloxane	Silica	Dispersion type	Temp (°C)	Polydimethylsiloxane on Silica Surface: Adsorbed (mg/m^2)	Polydimethylsiloxane on Silica Surface: Molecules 1000 $Å^2$
DC-MSA centrifugate	M-5	β	15.0 ± 0.2	4.9 ± 0.2	1.12 ± 0.05
	M-5	α	25.0 ± 0.2	0.50 ± 0.02	0.12 ± 0.005
	M-5	β	25.0 ± 0.2	4.9 ± 0.2	1.12 ± 0.05
	M-5	β	35.0 ± 0.2	4.7 ± 0.2	1.06 ± 0.05
	M-5	β	45.0 ± 0.2	4.9 ± 0.2	1.14 ± 0.05
DC-M centrifugate	M-5	β	15.0 ± 0.2	5.9 ± 0.2	1.42 ± 0.05
	M-5	α	25.0 ± 0.2	0.57 ± 0.02	0.14 ± 0.005
	M-5	β	25.0 ± 0.2	6.4 ± 0.2	1.53 ± 0.05
	M-5	β	35.0 ± 0.2	5.9 ± 0.2	1.42 ± 0.05
	M-5	β	45.0 ± 0.2	6.2 ± 0.2	1.49 ± 0.05
	G23	β	25 0 ± 0.2	2.5 ± 0.2	0.59 ± 0.05
	G35	β	25.0 ± 0.2	0.6 ± 0.2	0.15 ± 0.05
	G200	β	25.0 ± 0.2	1.6 ± 0.2	0.38 ± 0.05
DC-200 (1000 cSt)	A200	β	25.0 ± 0.2	4.2 ± 0.2	1.01 ± 0.05
Commercial DC-MSA		β	15.0 ± 0.2	3.1 ± 0.2	0.71 ± 0.05
		β	25.0 ± 0.2	3.5 ± 0.2	0.81 ± 0.05
		β	35.0 ± 0.2	3.4 ± 0.2	0.78 ± 0.05
		β	45.0 ± 0.2	3.4 ± 0.2	0.80 ± 0.05
Commercial DC-M		β	15.0 ± 0.2	3.2 ± 0.2	0.76 ± 0.05
		β	25.0 ± 0.2	3.5 ± 0.2	0.85 ± 0.05
		β	35.0 ± 0.2	3.5 ± 0.2	0.83 ± 0.05
		β	45.0 ± 0.2	3.0 ± 0.2	0.73 ± 0.05

[a] Sample calculation: (DC-MSA centrifugate + M-5, 15°C). DC-MSA centrifugate occupies 1.526 m^2/mg of polydimethylsiloxane on water surface. DC-MSA, M-5 dispersion occupies 0.887 m^2/mg of polydimethylsiloxane on water surface. 0.887 m^2 = area of spread polydimethylsiloxane + cross sectional area of silica.

1. Cross sectional area of silica = $n\pi r^2$. n = number of silica particles; since $n(4/3\pi r^3)(2.2\ g/cm^3) = 0.421$ mg = total weight of silica on water surface; $r = 6 \times 10^{-7}$ cm; $n = 2.64 \times 10^{13}$; $n\pi r^2 = 0.010\ m^2$.

2. Area of spread polydimethylsiloxane = 0.887 - 0.010 = 0.877 m^2; amount of bound polydimethylsiloxane = 1.526 - 0.877 = 0.649 m^2 or, 0.649 m^2/(1.526 m^2/mg) = 0.425 mg. The DC-MSA, M-5 dispersion contained29.63% M-5 silica, so per milligram of polydimethylsiloxane there is 0.421 mg of M-5; surface area of silica = 208 m^2/g $(0.421 \times 10^{-3}$ g) = 0.0876 m^2.

mg adsorbed polydimethylsiloxane/m^2 silica surface = 0.425/0.0876 = 4.9.

Since molecular weight of DC-MSA centrifugate = 2.6×10^4, there are 1.12 molecules of polydimethylsiloxane/ 1000 Å^2 of silica surface.

RESULTS AND CONCLUSIONS

Qualitative Tests

The following hypotheses have been proposed to explain the action of the added silica in enhancing foam inhibition by polydimethylsiloxane:

(1) silica prevents solubilization of polydimethylsiloxane by the solute in the foaming medium; (5)
(2) silica increases the dispersibility of polydimethylsiloxane in foams; (6)
(3) silica increases the interfacial area between polydimethylsiloxane and the foaming medium; (7)
(4) dispersions of silica in polydimethylsiloxane have an increased spreading coefficient; (8)
(5) silica increases the viscosity of the antifoam, which increases foam inhibition; and
(6) silica lowers the surface viscosity of spread films of foam inhibitor.

To test these hypotheses, we performed some simple qualitative experiments using commercial antifoams (CD-MSA and DC-M), and polydimethylsiloxane recovered by centrifugation from them.

Hypothesis (1) attributes the mediocre antifoam effect of polydimethylsiloxane alone to its being solubilized by the foaming medium; silica acts to prevent solubilization and so increases foam-inhibitory ability. This supposition is easily tested by adding polydimethylsiloxane in excess of the amount solubilized. No improvement in foam inhibition occurs when polydimethylsiloxane is added in excess of this amount. Therefore, while silica may retard or prevent solubilization of polydimethylsiloxane, such an effect is not central to its action in foam inhibition.

Defoaming by spreading of an antifoam is a well-known mechanism. (9) When polydimethylsiloxane is added to an aqueous foam, it will spread, but film rupture is not pronounced and hence defoaming does not occur to any marked extent. If more polydimethylsiloxane is added to the remaining foam, there is no additional film rupture. The remaining foam becomes heavily loaded with polydimethylsiloxane, which coats the lamellae. Since the polydimethylsiloxane is amply accessible to the foam even without silica, and since this accessibility anyway does not improve the foam inhibition, hypotheses (1), (2), and (3) will not hold. The immediate collapse of foam when a dispersed silica in polydimethylsiloxane is added implies the action of another mechanism for defoaming.

Addition of silica to polydimethylsiloxane does not change the spreading coefficient on water, (8) so hypothesis (4) is not a satisfactory explanation of the effect of added silica.

Table IV

Surface Viscosity of Films of Polydimethylsiloxane on Water

Spread Layer	Surface Concentration of Polydimethylsiloxane (mg/M^2)	Surface Viscosity (Surface Poise)
Polydimethylsiloxane Centrifugate from		
DC-MSA	0.50	1.0×10^{-4}
"	0.71	"
"	1.00	"
"	5.00	"
"	50.00	"
DC-MSA	0.50	1.0×10^{-4}
"	0.71	"
"	1.00	"
"	5.00	"
"	50.00	"

Varying the viscosity of polydimethylsiloxanes does not affect their (weak) foam inhibitory action for aqueous systems (8, 10), so hypothesis (5) also is not satisfactory.

Hypothesis (6) suggests that a spread film of polydimethylsiloxane helps to stabilize liquid lamellae because of a high surface shear viscosity. This explanation appears to be unaware that the surface shear viscosity of polydimethylsiloxane is actually low (less than 10^{-5} poise). (11) Nevertheless we measured surface viscosities of films of polydimethylsiloxane and a silica dispersion in polydimethylsiloxane (DC-MSA) spread on water with the surface-traction viscosimeter (see Table IV). (12) No difference in surface viscosity between spread films of polydimethylsiloxane and polydimethylsiloxane films containing silica was observed. These results demonstrate that hypothesis (6) is untenable.

Single Bubble Stability Tests

Measuring average lifetimes of single bubbles under a monolayer is a known technique. (13) Trapeznikoff (14) showed that polydimethylsiloxane spread on water actually stabilizes a water lamella. A monolayer or multilayers of polydimethylsiloxane on water stabilize bubbles blown under them in a manner similar to the bilayers of Trapeznikoff (see Figure 1). More significantly, a film of the α-dispersion on water also stabilizes bubbles, whereas a film of the β-dispersion does not. The similarity in behavior of bubbles created under a spread film of polydimethylsiloxane and a spread film of α-dispersion, and the dramatic difference introduced when the β-dispersion is substituted for either of them demonstrates the strong destabilizing effect produced by the hydrophobic silica particles of

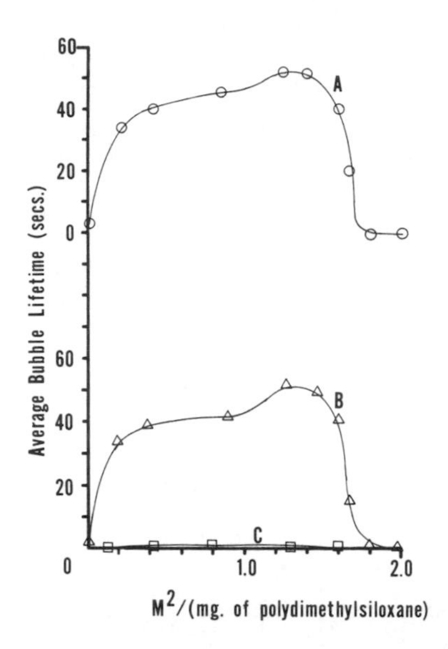

Figure 1 Average bubble lifetimes
A = Bubbles stabilized by polydimethylsiloxane, P (centrifugate from DC-MSA);
B = Bubbles stabilized by an alpha dispersion (Cabosil M5 in P; 6.03% solids);
C = Bubbles stabilized by a beta-dispersion (Cabosil M5 in P; 6.16% solids).

the β-dispersion in the polydimethylsiloxane film. This effect of the hydrophobic silica in a spread film is not taken into account by hypotheses (1) to (6).

Similar tests were attempted for films of polydimethylsiloxane, α-, and β-dispersions on aqueous solutions of sodium lauryl sulfate and hexadecyltrimethylammonium bromide. In both cases, β-dispersion destroyed bubbles instantly. For the other two agents, the results were erratic: bubble lifetimes are so long with those detergent solutions that solubilization of the monolayer and accidental contamination of the surface are major problems. However, one important conclusion was apparent. Single bubbles made with solutions of sodium lauryl sulfate and of hexadecyltrimethylammonium bromide last longer than five minutes, and so do the bubbles made when these solutions are covered with a spread film of polydimethylsiloxane or α-dispersion. Since a spread layer of polydimethylsiloxane or of α-dispersion has no destructive effect on bubbles, we conclude that polydimethylsiloxane does not defoam by replacing adsorbed solute at the surface as has been suggested. (15)

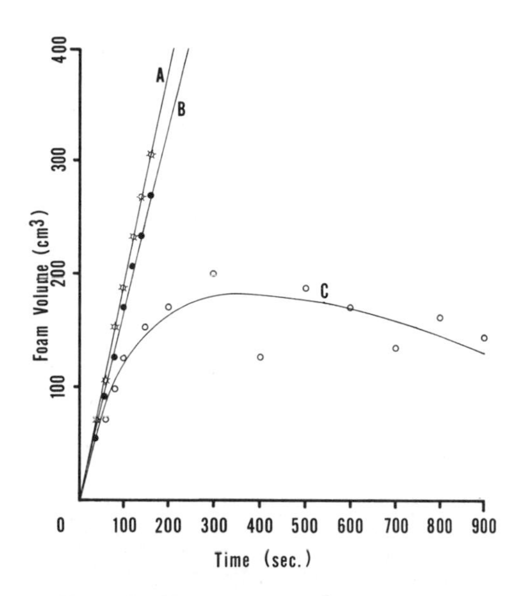

Figure 2 Dynamic Foam tests, foam volume versus time.
A = 0.02% sodium lauryl sulfate;
B = 100 ppm polydimethylsiloxane (centrifugate from DC-MSA) in A;
C = 100 ppm beta-dispersion (DC-MSA) in A.

Dynamic Foam Tests

The effects of low concentrations of polydimethylsiloxane, and of α- and β-dispersions on a foaming solution were measured with a dynamic foam meter. (3) Effective foam inhibition is possible only with β-dispersions (see Figure 2). No differences in foam inhibition could be detected by this test for β-dispersions made with different silicas.

Ross-Miles Tests

The relative effectiveness of foam inhibitors can be measured only when they are partially ineffective, that is, when the amounts added are insufficient for complete inhibition of foam. If only 0.6 ppm by weight of foam inhibitor is used in the Ross-Miles test, the effects of type and concentration of silica in the β-dispersion can be examined. Figures 3 and 4 report the results of such comparisons. Polydimethylsiloxane alone shows a slight inhibitory effect; when it contains β-dispersed silica it is more effective, and its effectiveness increases with increasing proportion of silica. The type of silica used is also important to the action of the foam in-

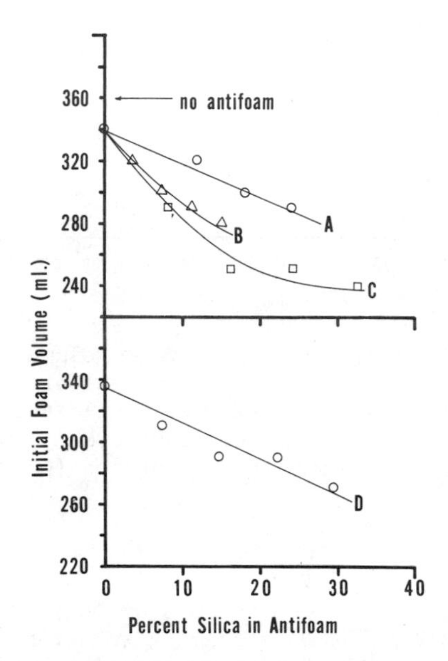

Figure 3 Ross-Miles foam tests, initial foam volume of a 0.400% aqueous sodium lauryl sulfate solution with 0.6 ppm antifoam.
A = Cabosil M5 in DC-M centrifugate as antifoam;
B = Gasil 35 in DC-M centrifugate as antifoam;
C = Gasil 23 in DC-M centrifugate as antifoam;
D = Cabosil M5 in DC-MSA centrifugate as antifoam.

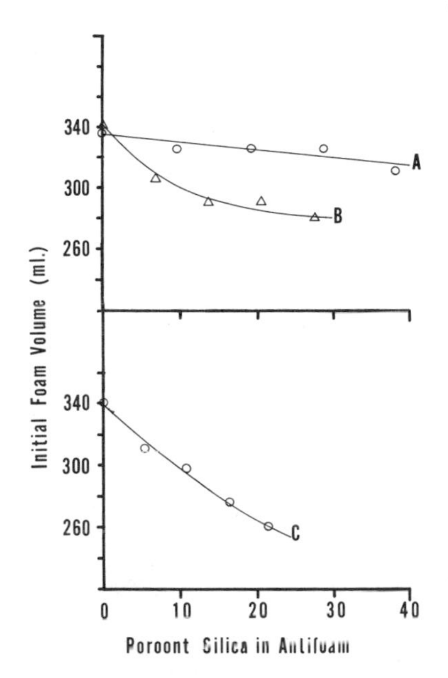

Figure 4 Ross-Miles foam tests, initial foam volume of a 0.400% aqueous sodium lauryl sulfate solution with 0.6 ppm antifoam.
A = commercial dispersion (DC-MSA) as antifoam;
B = commercial dispersion (DC-M) as antifoam;
C = Aerosil 200 in DC-200 as antifoam.

hibitor. In general, the larger-sized silicas are more effective additives for foam inhibitors (see Table I). Dispersions containing silica particles of diameter about 3 micrometers (G23, G35) are more effective than dispersions containing silicas of diameter about 0.1 micrometer (commercial disperions, M-5, A200).

The Dependence of Antifoam Effectiveness on the Degree of Hydrophoby of the Silica Surface

An improtant study of the fundamental considerations of antifoam activity was reported by Lichtman, Sinka, and Evans (5) of the Diamond Shamrock Company. This Company specializes in providing antifoams, in particular for the paper-manufacturing industry, and it markets formulations consisting of hydrophobic silica dispersed in hydrocarbon oil. The silica is made hydrophobic by treatment with polydimethylsiloxane before it is dispersed in the oil. (16) The experimental measurements reported by these authors, to which we would draw attention, are described in the following passage, translated from the Spanish of the original publication, and in Figure 5;

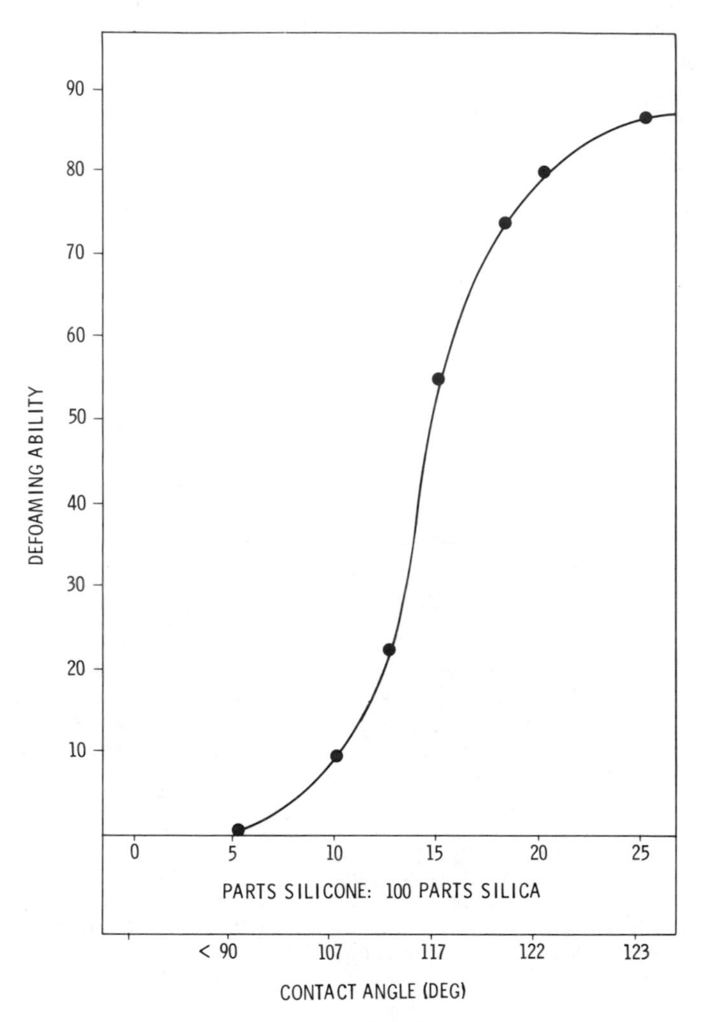

Figure 5 Contact angle of water on silica versus antifoam ability. (from reference 5, by permission)

"Particles of varying degrees of hydrophoby were prepared by treatment of silica with polydimethylsiloxane. (16) Antifoam formulations were prepared consisting of 10% hydrophobic silica dispersed in Gulf 333 Paraffin Oil. As shown in the Figure the contact angle of water increases as the content of the silicone oil increases, the particle becoming hydrophobic at about 5 parts of silicone oil to 100 parts of silica. Note that 'defoaming ability' becomes detectable only after the contact angle exceeds 90°.

'Defoaming ability' improves with enhanced hydrophoby up to 25 parts of silicone oil to 100 parts silica. The silica surface is saturated at this point and further addition of silicone oil would provide a product containing free silicone oil."

As the degree of hydrophoby increases, as measured by the increasing angle of contact, so does the vigor of the dewetting action, and concomitantly with it the decline in the adhesion of the medium to the particle by the aqueous medium. We conclude also that the α-dispersion contains silica whose surface, not being completely saturated with the bound polydimethylsiloxane, is not hydrophobic, which is manifested by its poor effectiveness as an antifoam; whereas the silica in the β-dispersion, being closer to saturated with polydimethylsiloxane, is strongly hydrophobic, and its effectiveness as an antifoam is fully developed.

The paper by Lichtman et al. also states that the hydrocarbon oil (Gulf 333 Paraffin Oil) has no antifoaming activity by itself in the absence of the hydrophobic silica. The antifoaming activity of the oil medium, when as in the case of silicone oil it has any, is less significant that the antifoaming activity of the dispersed hydrophobic silica; its function is to disperse the particles and prevent their flocculation, which would certainly occur were they in direct contact with an aqueous medium. It is sometimes stated that the function of the silica is to act as a dispersing agent for the silicone oil, but that statement misidentifies the principal actor in the antifoaming action.

The Effects of Dispersed Silica on the Antifoaming Properties of Polydimethylsiloxane

Our experimental results agree with the reports published elsewhere, especially in the patent literature, that the foam-inhibiting properties of polydimethylsiloxane are greatly improved, particularly with respect to "fatigue", by the presence of dispersed silica in its β-dispersion form. Silica is not the only suitable solid material for this purpose. The essential requirements are that the particles be hydrophobic (i.e., have a contact angle larger than 90°), that the dispersion be stable, and that the particle size lie between 0.2 and 5 μm. The solids may be naturally hydrophobic, e.g., talc, graphite, poly-α-olefin, ethylene or propylene homo- or copolymers, nylon, or polystyrene; (17) or they may be hydrophilic solids that have been treated to make their surfaces hydrophobic. (18) Examples of the latter are silica, bentonite, diatomaceous earth, attapulgite, and titanium dioxide. The immersion and heat treatment of the silica in polydimethylsiloxane is but one example of a treatment by means of which a polar solid can be made hydrophobic. Hydrophobic particles are readily dispersed in a non-polar medium.

The mechanism by which dispersed solid particles are so effective in destroying bubbles depends on the degree of hydrophoby of the particle. An oily surface causes a film of water to recede from it, and in the same way the surface of a hydrophobic solid particle causes water to recede, or to dewet, where the two make contact. The withdrawal of the water of the bubble lamella from the surface of the particle may by itself be enough of a mechanical shock to rupture the lamella and release the entrapped gas, or the rupture may be due to the poor adhesion between the water and the hydrophobic particle causing the water to fall away from the particle as it withdraws. Roberts _et al_. have published high-speed photographs that seem to demonstrate this mechanism for soap solutions defoamed by insoluble droplets. (15) Other demonstrations of lamella rupture by dewetting are the breaking of bubbles by sprinkling powdered talc on them, or by the mere touch of a finger, which is naturally hydrophobic, or of a Teflon-coated glass rod. The mechanism of rupture of liquid lamellae that we designate "dewetting" is equivalent in all respects to the "entering" of Robinson and Woods (20) and the negative of the spreading coefficient of the medium on the antifoam is identical with their "entering coefficient."

As we have shown, the size of the solid particles is important; Ross (19) had previously emphasized the importance of the droplet size in foam inhibitors. The larger particles or droplets, which are about the same size as the thickness of the liquid lamellae, are the more effective. The particles should be large enough to create sufficient motion of the liquid during dewetting to rupture the lamellae. The solid particles are an alternative to dispersed droplets of insoluble liquid; they have the advantage that the vigor of the dewetting action is not dissipated by the loss of droplet volume caused by spreading. The presence of the solid assures an ultimate particle size that cannot be attenuated. The chief function of the added hydrophobic solid is to ensure dewetting of the particle or droplet by the medium, while preventing or minimizing the spreading of the insoluble liquid.

Silicone fluids do not spread vigorously but they form stable spread films on water at any thickness. This is an unusual property for foam inhibitors, most of which are autophobic, i.e., they retract after spreading, leaving an oriented monolayer on water in equilibrium with a thick lens of bulk liquid. The monolayer is volatile, even for a high-boiling liquid such as tributyl phosphate. The cycle of spreading, retraction and evaporation of the monolayer can therefore occur repeatedly, causing violent agitation of the antifoam droplet on the foam lamellae. The spreading of the droplet is therefore an important mechanism for most antifoaming agents, silicones and perhaps other surface-active polymers excepted.

Polydimethylsiloxane is used without added silica for foam inhibition of lubricating oils and other organic liquids. (1)

Probably in these systems, in which the surface tension is low, polydimethylsiloxane would not spread on the foaming medium but breaks the bubbles by the dewetting mechanism. A similar situation occurs with aqueous solutions of perfluorinated solutes where the surface tension is reduced below 2×10^{-2} n/m. In such solutes the polydimethylsiloxane does not spread, and therefore acts as an excellent foam inhibitor by virtue of the dewetting mechanism, added silica particles not being required. Ultimately, however, if concentrations of surface-active solutes are increased to the point where the emulsification of the foam inhibitor takes place, the droplets or particles become too well stabilized to be dewetted and they cease to function as antifoams. Kulkarni, Goddard, and Kenner (21) have shown that this occurs at concentrations a little greater than the critical micelle concentration of the surface-active solute.

References

1. J. W. McBain, S. Ross, A. P. Brady, J. V. Robinson, I. M. Abrams, R. C. Thorburn, and C. G. Lindquist, "Foaming of Aircraft Engine Oils as a Problem in Colloid Chemistry", N.A.C.A. Wartime Report ARR No. 4105, Washington, D.C. (1944). The designations BO-1 and BO-2 are wartime pseudonyms for polydimethylsiloxane fluids.
2. S. Ross and G. Nishioka, Abstracts, 51st National Colloid Symposium, Buffalo, New York (1977). J. Colloid Interface Sci. (1978) In the press.
3. S. Ross and G. Nishioka, J. Phys. Chem., 79, 1561 (1975).
4. "1975 Annual Book of ASTM Standards", D 1173-53, pt. 30, p.180.
5. I. A. Lichtman, J. V. Sinka and D. W. Evans, Assoc. Mex. Tec. Ind. Celul. Pap. (Bol), 15, 26-32 (1975).
6. S. Ross and G. J. Young, Ind. Eng. Chem., 43, 2520 (1951).
7. J. E. Carless, J. B. Stenlake, and W. D. Williams, J. Pharm. Pharmac., 25, 849 (1973).
8. M. J. Povich, A. I. Ch. E. J., 21, 1016-1017 (1975).
9. S. Ross, Chem. Eng. Prog., 63, (9), 41 (1967).
10. L. A. Rauner, U. S. Patent 3,455,839; July 15, 1969.
11. N. L. Jarvis, J. Phys. Chem., 70, 3027 (1966).
12. J. T. Davies and G. R. A. Mayers, Trans. Faraday Soc., 56, 690 (1960).
13. W. B. Hardy, Proc. Roy. Soc., A86, 610 (1912); Collected Scientific Papers, 524-8 (1936).
14. A. A. Trapeznikoff and L. V. Chasovnikova, Kolloid Zh., 35, 990 (1973); Colloid J. of the USSR, 35, 926 (1973).

15. K. Roberts, C. Axberg, and R. Osterlund in "Foams", R. J. Akers, editor, Academic Press, London, 1976, page 39.

16. R. Liebling and N. M. Canaris, U. S. Patent No. 3,207,698; September 21, 1965.

17. F. J. Boylan, South African Patent 68-05,793; February 10, 1969; C. A. 72, 4711 (1970).

18. F. J. Boylan, U. S. Patent 3,408,306; October 29, 1968.

19. S. Ross, Rensselaer Polytechnic Institute Bulletin, Eng. Sci. Ser., 63, 39 (1950).

20. J. V. Robinson and W. W. Woods, J. Soc. Chem. Ind., 67, 361 (1948).

21. R. D. Kulkarni, E. D. Goddard, and B. Kanner, J. Coll. Interface Sci., 59, 468 (1977).

DISCUSSION

E. D. Goddard (Union Carbide Corp)

Can you comment on the nature of the bonding of silicone oil to the silica in the beta dispersions?

Author The irreversibility of the adsorption of the polymer when in the beta mode of bonding suggests something more than the relatively weak interaction that is commonly thought of as hydrogen bonding. Beta bonding probably takes place as an etherification of some silanol groups of the silica surface by means of the terminal hydroxyls of the polymer, with the elimination of water. The lack of change in the molecular weight distribution of the polymer after the reaction (unpublished results of our own) indicates that little or no transetherification, involving chain cleavage, occurs. The very nature of the materials in question makes it impossible to detect the formation of a covalent bond, because so many Si-O bonds are already present in both reactants. Yet the formation of such a bond cannot be ruled out: the existence of strong hydrogen bonds does not preclude the presence of covalent bonding as well.

M. N. Yudenfreund (Drew Chemical Corp)

Have you determined the particle size distribution of silica in these dispersions? Would you comment on the relationship between the particle size distribution and defoamer efficacy?

Author We have not determined the particle size distribution of the silicas that we used in this study nor do we see much point in doing so. The larger particles are so much more effective than the smaller ones that the top end of the distribution would do all the work of defoaming and so leave the rest of the particles in the distribution without the opportunity to demonstrate whatever effect they might have.

M. N. Yudenfreund (Drew Chemical Corp)

Can one quantize the defoaming efficacy of a particular dispersion by determing the ratio of $\beta:\alpha$ by means of hexane induced precipitation? Have you done this?

Author We have not done this. It should be possible to do so, but there are simpler and more direct ways to quantize the defoaming effectiveness of a particular dispersion.

E. D. Goddard (Union Carbide Corp)

The authors are to be congratulated on debunking the very faulty explanations given in the literature concerning the role of hydrophobic silica in silicone oil in defoaming. Results in our laboratory, as far as they have gone, agree essentially with their basic mechanism but we differ slightly as regards the details of the mechanism we have proposed. It includes adsorption of surfactant by the particles once they have been carried by the silicone carrier oil into the lamellar interfaces.

Author Not all foams are stabilized by the same kind of solute and so this suggested mechanism of adsorption of solute by the silica seems rather special. Hydrophobic silica destroys a liquid film of water that would otherwise be stabilized by means of a monolayer of polydimethylsiloxane. Adsorption of the stabilizer by the silica is ruled out in this case. If the particle works there by causing dewetting, why not elsewhere by the same mechanism, which is perfectly general and applicable to all solutes and solvents.

R. D. Kulkarni (Union Carbide Corp)

Please comment on the degree of silicone oil adsorption in relation to the surface properties of silicas used (especially since you have used different types of silicas in this study).

Author Cab-O-Sil M-5 and Aerosil 200 are produced by flame hydrolysis of silicon tetrachloride in the gas phase at 1100 C. They are nonporous hydrophilic silicas. Gasil silicas are described as "micronised silica gels" and consist of larger and porous particles. Since the concentration of surface silanol groups on all these silicas is higher than the number that actually react with polydimethylsiloxane (3 per 100 sq. A. versus 1 per 1000 sq. A) we do not believe that differences in surface-chemical characteristics can significantly affect the amount adsorbed. The porous nature of the Gasil silicas, however, seems to exclude a lot of surface from adsorption.

E. J. Clayfield (Shell Research Ltd, England)

Could you clarify one point mentioned in your explanation of antifoaming mechanism, namely, the relative ineffectiveness of silicone oil alone being attributed to the tendency of a droplet of this material to spread at the foam lamella surface. How can material of such relatively low spreading pressure (8 dyne/cm from your results) spread over a surfactant-covered foam lamella surface of low surface tension?

Author Polydimethylsiloxane spreads on water with a relatively low spreading pressure because the interfacial tension against water is high (about 45 dynes/cm). But it also spreads on aqueous solutions of most surface-active solutes, which must be the result of a considerably lower interfacial tension against such solutions.

R. Kulkarni (Union Carbide Corp)

Could you comment on as to how can insure a certain antifoam droplet size (>2 μm size) by incorporation of β dispersed silica especially since TLO silica size is of the order of 500 Å or so. (One can indeed produce >2 μm silicone oil droplets even without the use of silica and yet it does not work as antifoam.)

Author We refer you to the publication on Cab-O-Sil issued by Cabot Corporation, describing its manufacture. The primary particles are indeed less than 200 Å in diameter, but they fuse with one another on cooling so that the end product, or aggregate, consists of particles that are a few tenths of a micro-meter in diameter. We consider these particles and small undissociated agglomerates of them, when coated with polydimethylsiloxane, to be the active ingredient in the antifoam; and such particles do guarantee a size equal to their diameter. Gasil silicas are precipitated silicas and are 3 to 5 micro-meters in diameter. These particles guarantee a larger size of antifoam particle and we find them to be a more effective additive than Cab-O-Sil M-5 or Aerosil 200. We agree that one can produce droplets of liquid polydimethylsiloxane of about 2 μm, but these are quickly dissipated whereas a single silica particle is not.

DIELECTRIC RELAXATIONS IN SHEARED ORGANO-BENTONITE DISPERSIONS

R. J. Kuo, R. J. Ruch, and R. R. Myers
Department of Chemistry
Kent State University
Kent, Ohio

I. INTRODUCTION

The mechanical properties of a colloidal dispersion depend largely on specific interactions between the disperson medium and the dispersed phase. These interactions, in turn, are responsible for the stability of the system as measured in such practical terms as sedimentation, flocculation, caking, color drift, and gloss.

The stability of a dispersion reflects its resistance to molecular or chemical disturbances [1], and is affected by such factors as temperature, concentration, shear and formulation variables such as order of addition. In many instances one desires a certain degree of instability, in recognition of the fact that dispersions are metastable compositions at best, so that the dispersed particles make a fragile continuum of their own which is readily dispersed by shear. Such a system is characterized by a large sediment volume [2].

Electrical charges at interfaces are bound to play a part in establishing sediment volumes and the various mechanical properties of dispersions, particularly in media of low dielectric constant; as a consequence, dielectric constant measurements were selected as a sensitive probe of the behavior of such dispersions under various degrees of structure breakdown under shear.

Lyklema [3] states that electrostatic repulsion is a minimal stabilizing factor for non-aqueous colloidal dispersions, unless there is a dissociation of surface groups or adsorption of ionic surfactants; frequently, a more significant steric factor called entropic stabilization or osmotic repulsion is believed to operate. This factor, first noted by van der Waarden [4] and explained by Mackor [5] and later by Fischer [6], is attributed to the presence of an adsorbed long-chain species on the surface of the particles comprising the dispersed phase.

The stability resulting from adsorbed polymer films has been studied extensively and is not the subject of this reasearch. The object of this paper is to describe a sensitive method for measuring dielectric constants ε of dispersions under controlled rates of shear, showing thereby that ε cannot be represented simply as a single-valued quantity and demonstrating the effect of interfacial treatments on the complex components, ε' and ε''.

As early as 1947, Voet [7] measured ε of non-aqueous dispersions of carbon black under shear. Relying on wide differences between ε of carbon black and that of the hydrocarbon medium, Voet extended an equation developed by Bruggeman [8] some ten years earlier by including in it a form factor to accomodate non-spherical particles. Flocculation naturally leads to non-spherical geometry, and as a consequence the form factor became a measure of the reticulations of the carbon particles.

The need to consider the dielectric constant as a complex quantity ε^* was expressed in a review by Dukhin [9] in which the effects of interfacial polarization on heterogeneous systems is discussed. Hanai [10] reviewed these effects in terms of their bearing on the two components, ε' and ε'', of ε^*, included his own extension [11] of the relevant theory and described the means employed by Debye [12], Cole and Cole [13], and Davidson and Cole [14] for graphical representation of the data.

A more general form for representing dielectric relaxation was proposed by Havriliak and Negami [15] and was termed the

Skewed Circular Arc Equation (SCAE). Under proper conditions, the SCAE reduces to the simpler semi-circular arc, the circular arc, or the skewed arc equation. In the simplest form known as the semi-circular arc:

$$\varepsilon^* = \varepsilon' - i\varepsilon'' = \varepsilon'_\infty + \frac{\varepsilon'_0 - \varepsilon'_\infty}{1 + i\omega\tau_0},$$

the complex dielectric constant, ε^*, reduces to ε'_0 at an angular frequency of $\omega = o$, and ε'_∞ at $\omega = \infty$, in which cases there is no relaxation component; that is, there is no contribution from the imaginary term containing τ_0, the relaxation time.

A plot of ε', the real part of ε^*, against ε'', the imaginary part, is called an Argand diagram or complex plane plot. When $\omega = 1/\tau_0$ for a semi-circular arc, a single relaxation time is located on the Argand diagram at $\varepsilon' = (\varepsilon'_0 + \varepsilon'_\infty)/2$ and $\varepsilon'' = (\varepsilon'_0 - \varepsilon'_\infty)/2$. In practice the Argand diagram is seldom semi-circular with its center on the ε' axis; the relaxation term takes the form $[1 + (i\omega\tau_0)^{1-A}]$ and/or its weight in the denominator is some quantity B less than unity.

Oesterle and Müller [16] reported that a difference between "effective wetting" and "psuedo wetting" could be determined between pigment and binder by dielectric measurements. The present investigation was undertaken to extend this concept to see if complex dielectric quantities would provide an insight into the interfacial properties of the dispersed system. In this study, both the particulate phase and the medium had low ε so as to emphasize interfacial effects. Bentonite clay was dispersed in an oligomer, polybutene, under conditions in which its surface was systematically ion-exchanged with a homologous series of organic amine salts so that its interfacial interactions with the vehicle would change in a systematic or predictable manner. It was anticipated that the dielectric properties of the colloidal dispersion could be related to the stabilizing factors in the system.

II. MATERIALS AND METHODS

A. Dispersion Medium

Polybutene (Indopol L-14) obtained from Amoco Chemicals was used as the dispersion medium. The polymer backbone resembles polyisobutylene with about 90% of the chains containing a terminal monoolefin unit. The material had a number average molecular weight of 320 and a dielectric constant ε of 2.27 with no dielectric loss over the range of 20 Hz to 100 KHz. That is: $\varepsilon^* = \varepsilon' = \varepsilon$.

B. Dispersed Phase

Centrifuged, spray-dried Wyoming bentonite was supplied by the Baroid Division of NL Industries. Labile cations (mostly Na^+) on the clay surface were exchanged with amine cations by reaction with selected amine hydrochlorides and subsequent purification by Soxhlet extraction with water and isopropanol. These organo-clay adducts were prepared from hexylamine (C-6), octylamine (C-8), decylamine (C-10), dodecylamine (C-12), tetradecylamine (C-14), and hexadecylamine (C-16).

The purified organo-bentonite dried cakes were pulverized in a mortar and pestle, passed through a 200-mesh sieve, vacuum dried, and then milled at various weight percentages with polybutene in a ceramic ball mill for about seven hours.

C. Rotary Cell

The stainless steel cup and bob of an Interchemical precision rotational viscometer were used as the two electrodes of the dielectric cell. The bob was insulated from the cup which was connected to ground through the viscometer. The empty cell capacitance was 22 pf. The cell could be rotated at a velocity that produced a maximum shear rate of 292 sec^{-1}.

D. Capacitance-Conductance Bridge

A General Radio 1610B-2 bridge assembly was used to measure the capacitance and dissipation factors of the dispersions in the frequency range of 20 Hz to 100 KHz. The substitution method was used. A stray capacitance of 2.7 pf. was established by calibration with standard liquids.

E. Temperature

A water bath surrounding the rotary cell was controlled to $\pm$ 0.05 C over a temperature range of 5 to 60 C by using a thermoregulator and a cooling probe.

F. Argand Diagrams

The Skewed Circular Arc Equation (SCAE): $\varepsilon^* = \varepsilon'_m + (\varepsilon'_0 - \varepsilon'_\infty)[1 + (i\omega\tau_u)^{1-A}]^{-B}$ was analyzed by the skewed circular arc method of analysis (SCAMA) [17]. A Hewlett-Packard 9820A calculator was used to adjust parameter A and B for the best fit to the experimental data.

III. EXPERIMENTAL RESULTS

A. Shear

The dielectric measurements of all of the colloidal systems studied were sensitive to the applied shear rate, $\dot{\gamma}$. The magnitude of tan δ_{max} (where tan $\delta = \varepsilon''/\varepsilon'$) decreased as the shear rate increased, although the frequency at which the maximum occurred remained fairly constant; e.g., just below 1 KHz as illustrated in Figure 1 for the C-10 system.

The breadth of the complex plane plots decreased as the shear rate $\dot{\gamma}$ increased. This is illustrated in Figure 2 for the C-10 system. It should be noted from the disappearance of the tails at

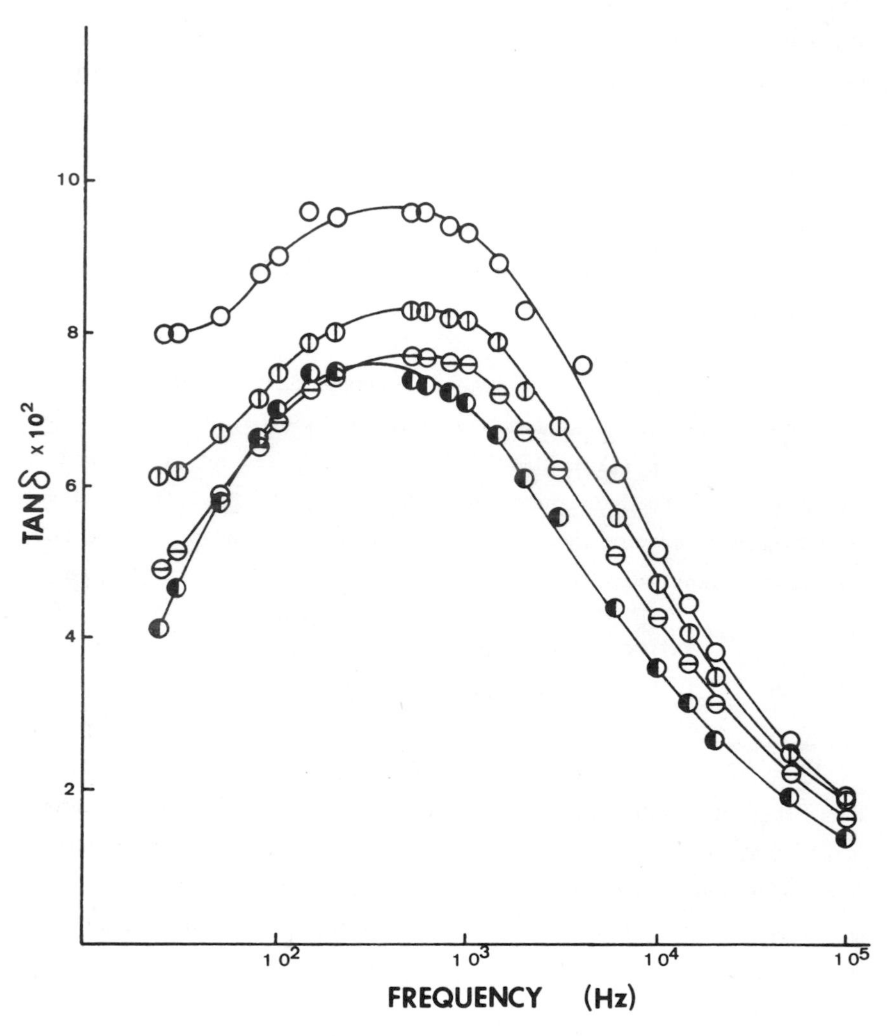

Figure 1 Frequency dependence of tan δ as a function of shear rate at 30 C for C-10 at 12% by weight. ○ - rest; ⦶ - 7.3 sec^{-1}; ⊖ - 73 sec^{-1}; ◐ - 292 sec^{-1}.

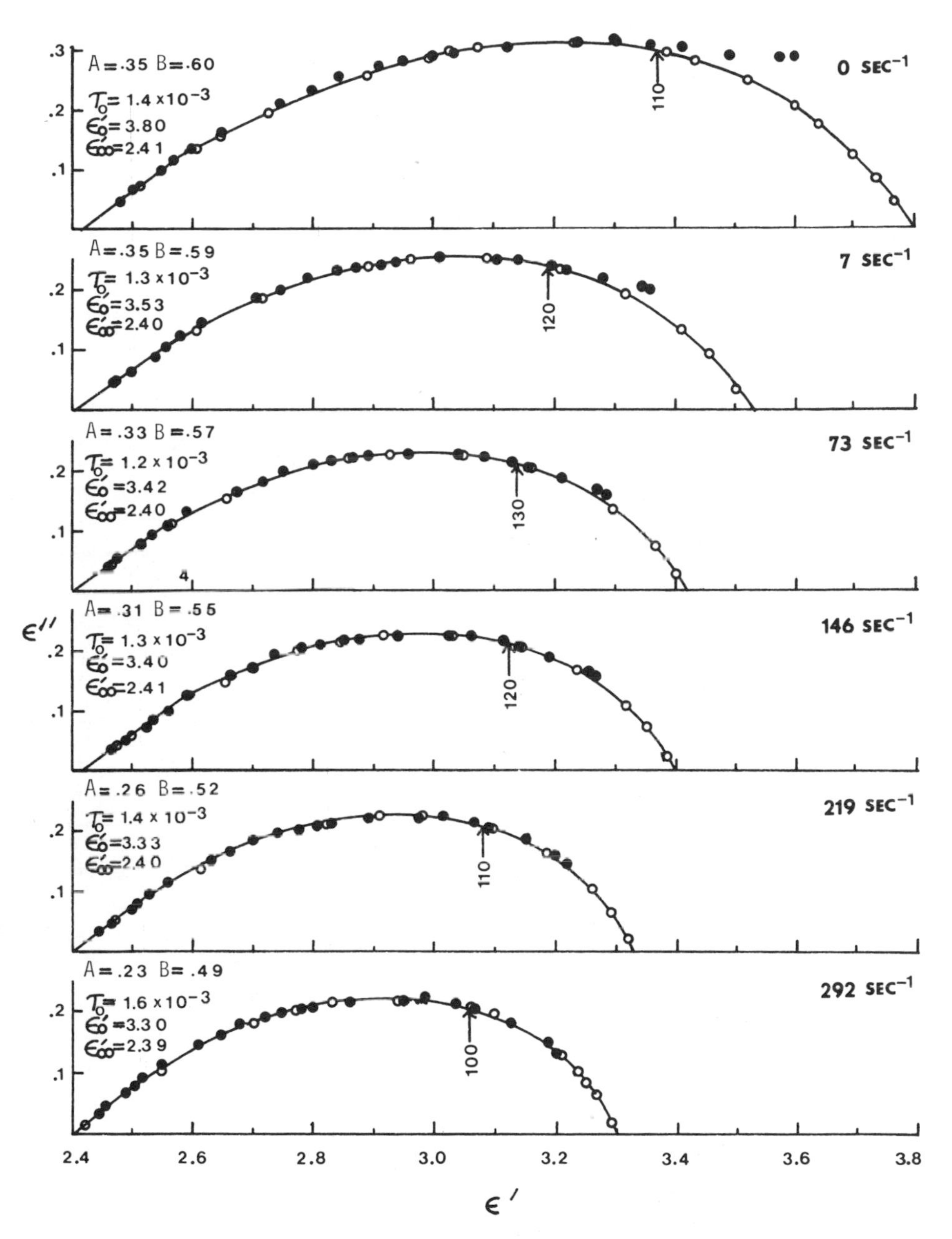

Figure 2 Complex plane plot of 12% by weight C-10 at 30 C from rest to a shear rate of 292 sec^{-1}. Solid circles represent experimental values. Open circules and the solid line are calculated by the SCAMA. Numbers preceding arrows are relaxation frequencies. Values in the upper left portion of each curve are the parameters of the SCAE.

higher $\dot{\gamma}$ that the deviation from the SCAE in the low frequency region decreased as $\dot{\gamma}$ increased. The relaxation time, τ_0, was relatively insensitive to $\dot{\gamma}$, remaining at 0.0014 $\pm$ 0.002 sec over the shear range shown. Except for the lowest shear rate, the values of parameters A, B, and ε'_0 decreased in a linear fashion as $\dot{\gamma}$ increased, while ε'_∞ remained essentially constant.

B. Aging

τ_0 values for the organo-clay adducts decreased with aging. C-6 and C-8 achieved a constant τ_0 within 50 days; C-10 and C-14 required 50 to 80 days, and τ_0 was still decreasing for C-12 and C-16 80 days after the collodial dispersions were prepared.

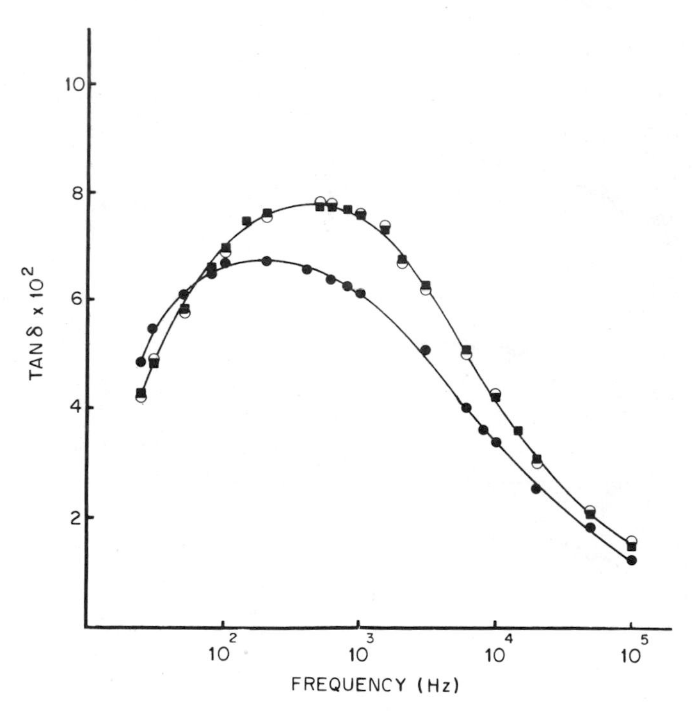

Figure 3 Frequency dependency of 12% by weight C-10 at 30 C and a shear rate of 292 sec^{-1} as a function of aging.
● - freshly prepared dispersion;
■ - 50 days aging;
○ - 80 days aging.

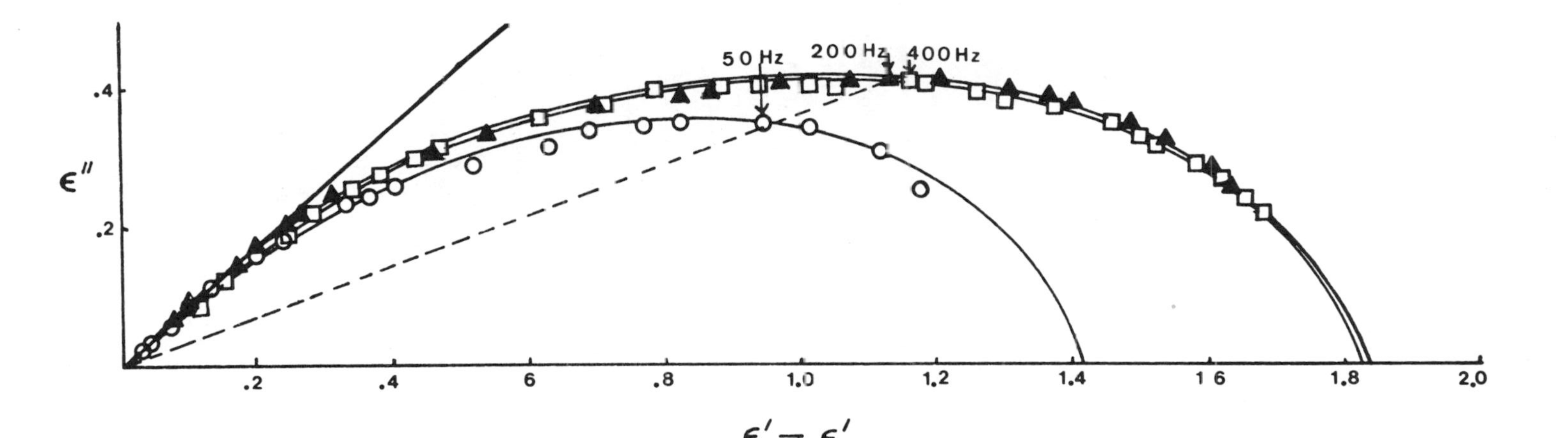

Figure 4 Complex plane plot of 12% by weight C-16 at 30 C and a shear rate of 292 sec^{-1} as a function of aging. The dotted line intersects the curves at the relaxation frequency. O - freshly prepared dispersion; ▲ - 50 days aging; □ - 80 days aging.

The magnitude of tan δ_{max} increased with aging for all of the systems, but in all cases, the maximum value occurred within 50 days. The frequency at which tan δ_{max} occurred shifted to higher values over the same time frame as the τ_o values changed. Figure 3 shows a tan δ_{max} versus frequency plot for C-10. Figure 4 for C-16 shows that even for the longer chain systems, the increase in the breadth of the dielectric dispersions occurred entirely within the first 50 days of aging.

Sedimentation volumes were determined on samples aged for about 100 days. A logarithmic plot of sedimentation volumes versus time gave the following negative slopes: 0.95 day^{-1} for the raw clay, 0.22 day^{-1} for C-6, 0.18 day^{-1} for C-8 and C-10, 0.08 day^{-1} for C-14, and 0.05 day^{-1} for C-12 and C-16.

C. Temperature

The individual numerical value of tan δ_{max} for each system studied did not change appreciably over the temperature range of 10-50 C. The frequency at which tan δ_{max} occurred did, however, increase as the temperature increased.

As the temperature increased, the fit to the SCAE in the low frequency range became poorer. It was also more difficult to determine ε'_∞ because the ε' values shifted to the low frequency side of the dispersion. The numerical values for A, B, and ε'_o increased in a linear fashion as the temperature increased for the C-6 system. ε'_∞ decreased linearly for C-6 as the temperature increased from 6 to 25 C permitting an extrapolation to determine the values for ε'_∞ at the higher temperatures. Figures 5 and 6 show the SCAE plots for C-6 over the temperature range of 6-30 C. The low frequency deviations from the SCAE were even more pronounced at 40 and 50 C.

A plot of log τ_o vs. 1/T for the C-6 system was linear over the temperature range of 6-50 C and gave a $\Delta E^{\ddagger}$ value of 20 kcal.. Similar plots for the other systems were not linear over the entire temperature range.

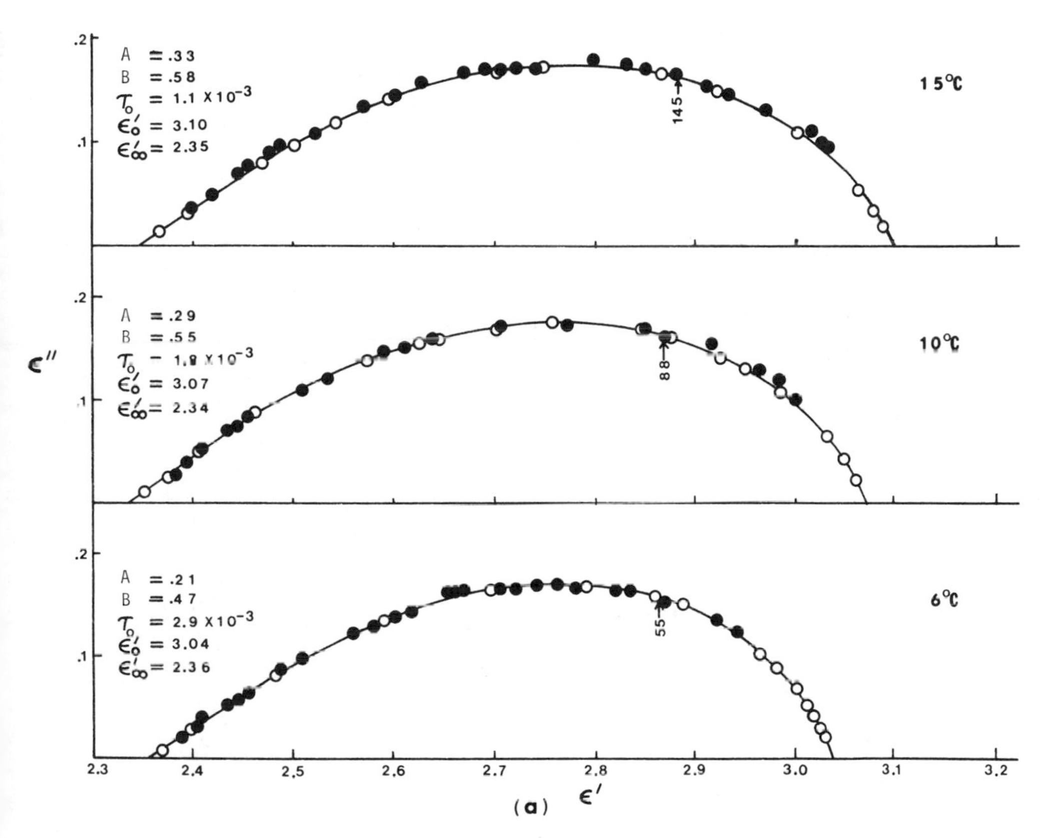

Figure 5 Complex plane plot of 12% by weight C-6 at a shear rate of 292 sec^{-1} and three temperatures. Symbols as previously defined.

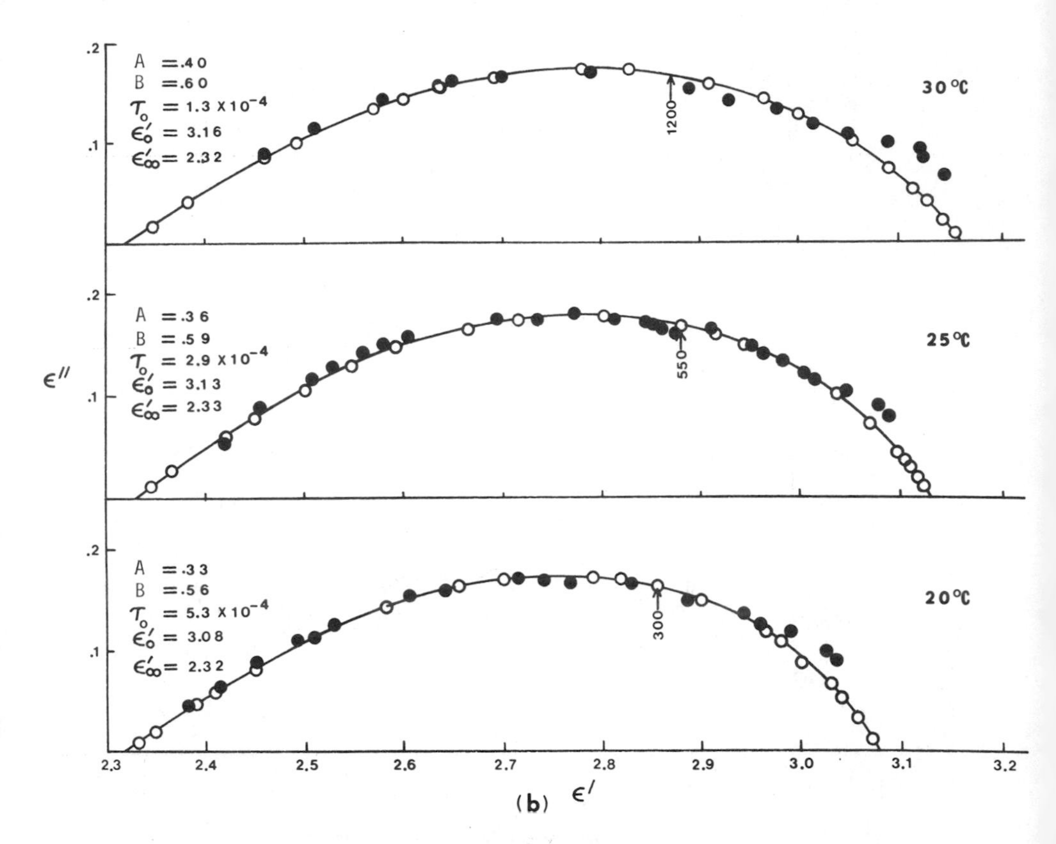

Figure 6 Complex plane plot of 12% by weight C-6 at a shear rate of 292 sec^{-1} and three temperatures. Symbols as previously defined.

D. Concentration

The C-6 system was investigated over the concentration range of 7% to 30% by weight. $\varepsilon'_o - \varepsilon'_\infty$ increased from 0.4 to approximately 2.8 over this concentration range at a shear rate of 292 sec^{-1}. ε'_∞ was difficult to determine accurately at weight concentrations above 14% due to a shift of the ε' values to the low frequency side of the Argand plot. Deviation from the SCAE also occurred in the low frequency range at the higher concentrations. The numerical value of tan δ_{max} increased from 0.02 to 0.14 over the concentration range although the frequency at which tan δ_{max} occurred did not change appreciably. Estimated values of τ_o

suggested there was also no appreciable change in its value over this concentration range.

E. Chain Length

In order to minimize low frequency conductivity effects, the highest shear rate of 292 sec^{-1} was chosen to make comparisons among the C-6 through C-16 systems. An aging of 50 days at 30 C and a weight concentration of 12% were chosen to give uniform conditions for comparison of the systems.

SCAE plots for the C-12, C-14, and C-16 systems under the stated conditions are shown in Figure 7. ε'_∞ did not change much from system to system and was near the 2.27 value for the

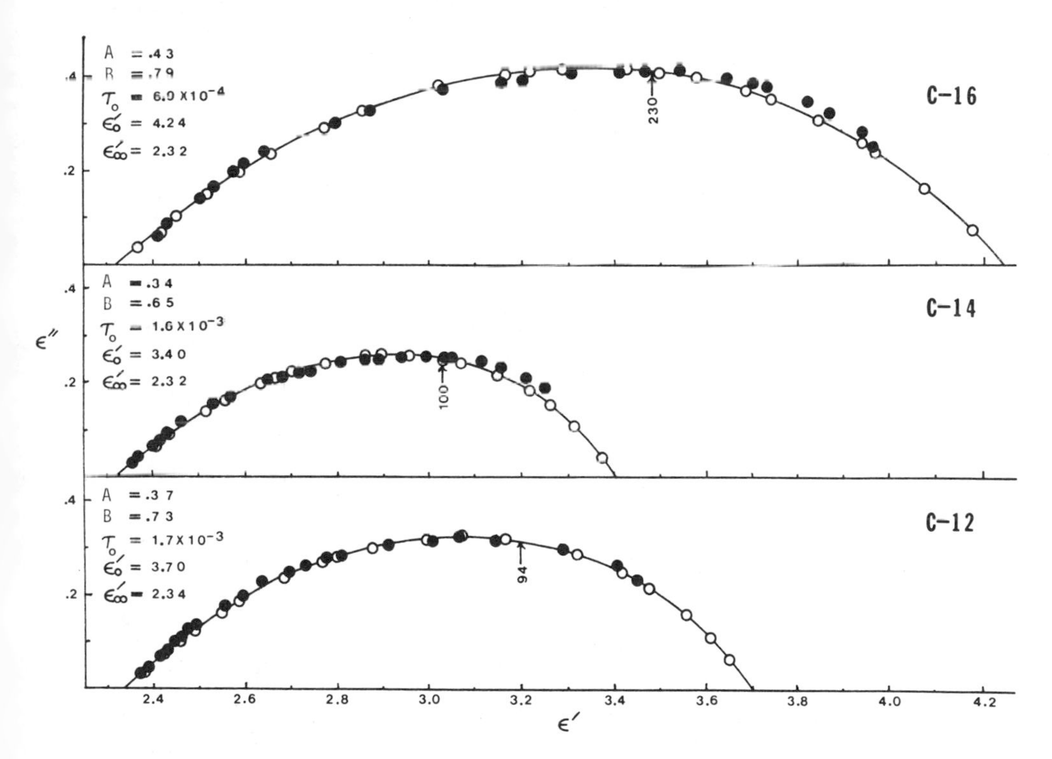

Figure 7 Complex plane plots of 12% by weight C-12, C-14, and C-16 at 30 C and a shear rate of 292 sec^{-1}. Symbols as previously defined.

polybutene dispersion medium. Allowing for the inversion involving the C-12 and C-14 systems, there was a fairly smooth transition in the numerical values of ε_0', A, and B with respect to the number of carbon atoms in the amine chains, although ε_0' and A were significantly higher for C-16 than for the other systems.

The tan δ_{max} and B values determined at 50 days correlated well with 100-day sediment volumes as shown in Figures 8 and 9. (The inversion of sedimentation volumes for C-12 and C-14 was also found by Jordan [18] with various vehicles and by Dawson and Haydon [19] for rutile particles coated with the corresponding carboxylic acids. No explanation for the inversion was offered in either study.) Tan δ_{max} for the raw clay (C-0) was the value measured at 100 KHz, the limit of the high frequency measurements.

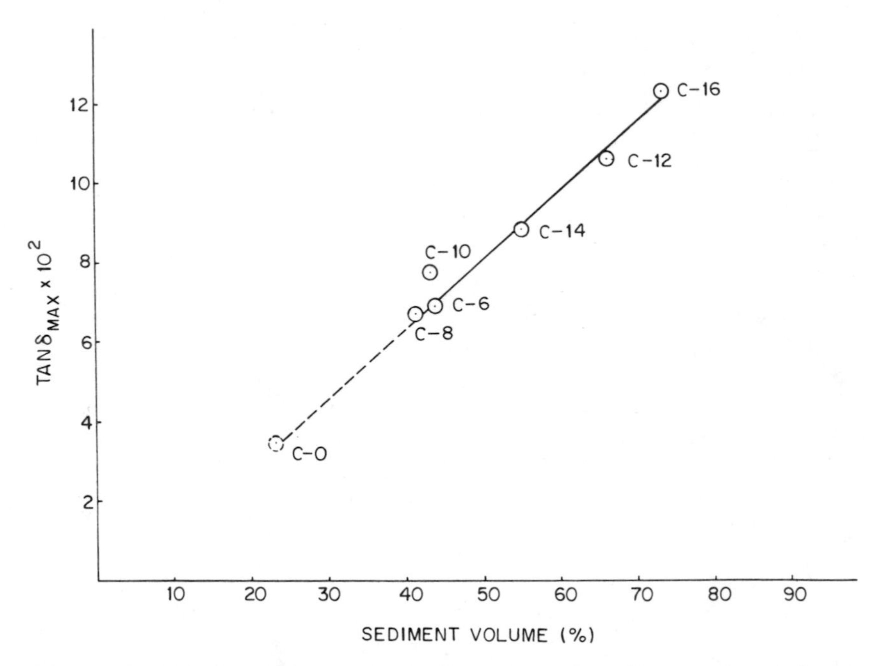

Figure 8 Tan δ_{max} plotted as a function of sedimentation volume for the raw clay and the C-6 through C-16 systems. The value for the raw clay, C-0, is at 100 KHz.

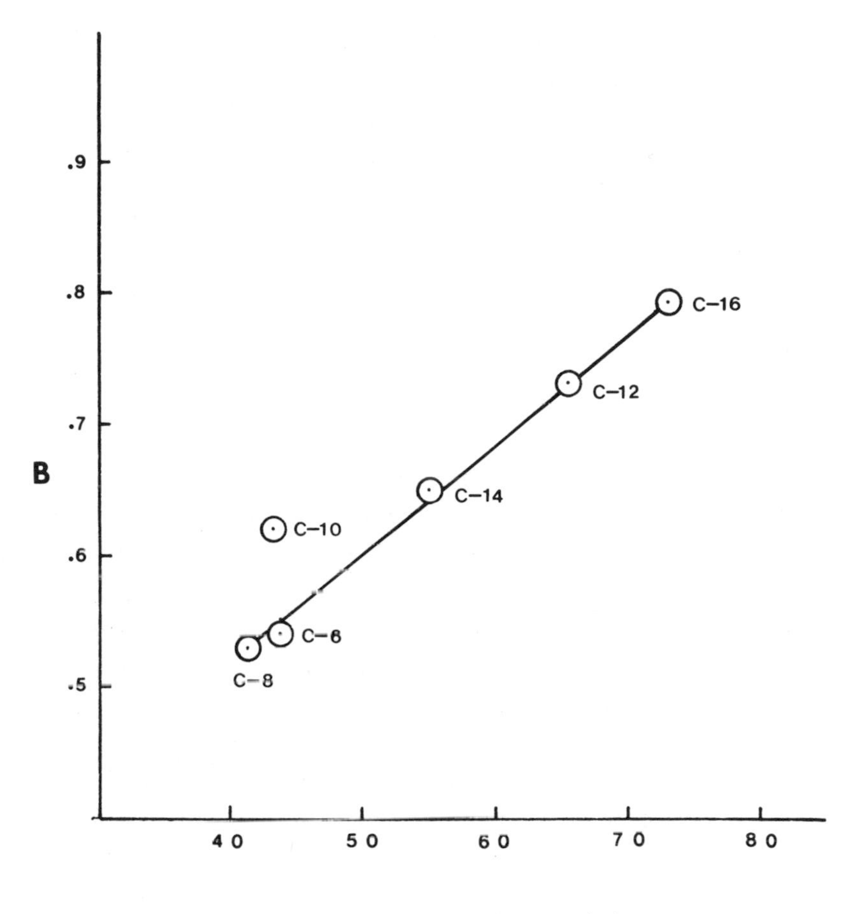

Figure 9 Parameter B of the SCAE as a function of sedimentation volume for the C-6 through C-16 systems.

IV. DISCUSSION

Dielectric relaxation processes can arise from the inability of polar molecules or groups to respond in phase with an applied electrical field (Debye) or from the inability of conductive charges to respond in phase with a changing electrical field (Maxwell-Wagner). Although Shah [20] has shown that bentonite has a permanent dipole moment, it is unlikely that the dielectric relaxations of the magnitude and frequency range found in this study arise from the oscillation of the coated clay particles.

In order to understand the source of the dielectric dispersion in the present study, it is necessary to describe the nature of the colloidal dispersion and the factors which affect it. The milling process reduces the coated clay particles to as near a division of fineness as possible by mechanical work. The particles in this condition are platelets [18] which are able to set up edge-to-edge and edge to face networks capable of dielectric loss as shown by the maximum in tan δ in Figure 1.

The deviations at low $\dot{\gamma}$ in Figure 2 emphasize the conductivity loss which is most pronounced in the low frequency range and is displayed only at or near rest conditions. Increasing shear breaks up the conducting chains, reducing the low-frequency D. C. conductivity and virtually eliminating it at the higher shear rates. This observation is reminiscent of that of Voet [7] who pointed out that conductive chains of carbon particles could be formed in collodial dispersions and that the number of chains is reduced by the application of shear; but the explanation cannot be found, as his was, in terms of the conductivity of the particles. Conduction in the case of the coated clay particles must involve the interface.

The decreasing breadth of the complex plane plot with increasing $\dot{\gamma}$ in Figure 2, stemming from the reduction of ε_o', reflects the structural breakdown capabilities of shear. In addition, the long axis of the coated clay particles is perpendicular to the applied field, which will also lead to a decrease in ε' as pointed out by Altshuller [21].

Because the primary effect of interfacial polarization would not be affected by a systematic increase in the chain length of the bound amines, the property differences must be due to secondary effects, the most prominent of which is reticulation of the particles. The increases in breadth of the Argand plots with increased temperature, concentration, chain length, and age all point to reticulation: temperature increase allows neighboring particles to penetrate the sheath, concentration increases proximity, chain lengthening improves compatibility with vehicle, and aging provides the time needed for the molecular displacements to take place. Dispersons are metastable entities at best, and the only stable morphology is the one in which the two phases have coalesced into independent masses with a single interface; therefore these various influences are the ones that tend toward ultimate stability.

The decrease in breadth of the Argand plots with shear augments the deductions reached above. Structures are destroyed by shear, and the effects of this destruction are best observed at low frequency, where the kinetic unit responding to the stress is largest. High-frequency measurements activate only segments of molecules and make semiconductors appear like insulators, so that responses in the vicinity of ε'_∞ are all likely to exclude interfacial polarization effects, including the secondary ones. The region near ε'_0 is the more sensitive one.

The frequencies at which the various curve parameters are found also reflect the behavior of a system with a polar bilayer continuum. The lossiness of the dispersions as reflected in tan δ_{max} increased in the same order as the increase in ε'_0 (which provided for the height of the curve on which tan δ_{max} is based). This correlation is to be expected; however the shift in frequency at which the loss peak is attained increased upon both heating and aging is not deducible from any *a priori* reasoning unless the reticulum model is invoked. Heating and aging both allow a perfection of the bilayer in that the meshing of the organic tails of adjacent particles is facilitated by time and by the reduction of viscosity with heating.

Jordan [18] points out that through C-10, less than half of the basal plane of the clay platelets is covered by the amine, permitting unhindered overlap of platelets with a resultant low basal plane spacing. C-12 and beyond have a higher basal plane spacing because the platelets have over 50% coverage. This situation helps explain the aging process and concurrent shifts of tan δ_{max} values to a higher frequency, even after the sediment volume has become constant. The shorter chain systems reach a minimum relaxation time in a short period of time, indicating that the particle-vehicle interactions are achieved fairly rapidly and that further particle breakdown is not very significant. Ball milling reduced the coated particles to aggregate size, after which penetration by vehicle further subdivided the aggregates to their ultimate particles in the case of C-12 and higher. The longer chain systems, with greater than 50% basal plane coverage, are open to basal plane penetration by the dispersion medium and dispersal into smaller platelets. Oesterle and Müller [16] observed a shift of tan δ_{max} to higher frequencies when grinding time was increased and postulated that the frequency of tan δ_{max} increases with a decrease in particle size.

With the increase in the number of particles in the dispersion medium, and the extension of more C-12, C-14, and C-16 chains into the polybutene vehicle, the chances increase for development of a structured system. To amplify this effect, an aged dispersion of Bentone 38, a dimethyldioctadecylamine derivative of bentonite, was compared with the prepared dispersions; it displayed a high dielectric loss in the low frequency range and subsequent formation of a gel, both of which indicate reticulation.

The sedimentation volume percentages of the aged C-6, C-8, and C-10 systems were 43 ± 2, suggesting that the bulk structures set up for these compounds with less than 50% basal plane coverage were of a similar nature. There would not be a strong driving force to separate these mechanically ground platelets since the amine chains can undergo lateral interaction within the layered clay structure. The cleavage of basal planes in C-12 and above would lead to an increase in entropy for the amine chains and there would be more extensive interactions of the amine chains with the dispersion medium than with one another within the interlayer regions of the clay platelets.

Mackor [5] has shown that the steric repulsion between colloidal particles with adsorbed hydrocarbon chains is due to the decrease in the number of possible chain configurations as the particles approach one another. In this study, the C-6 through C-10 systems would suffer the greater reduction in entropy on approach since these surfaces have less than 50% coverage and a packing separation of one hydrocarbon chain, or 4 angstroms. The corresponding separation for C-12 through C-16 is 8 angstroms because two chain widths have to be accommodated when coverage exceeds 50%. As a result, the C-6 through C-10 systems do not agglomerate, but the individual particles settle to give low sedimentation volumes. The C-12 through C-16 systems encounter primarily steric repulsion and not the degree of entropic repulsion found in the shorter chain systems; as a consequence they flocculate and give higher sedimentation volumes.

At a constant shear rate, an increase in temperature leads to an increase in collisions of the particles and an increase in conductivity in the low frequency range as shown in Figures 5 and 6. The breadth of the dispersion also increases with an increase in temperature due to the more random orientations of the particles, as predicted by Altshuller [21]. The increase in temperature also weakens the boundary layer interactions of the amine chains with the dispersion medium and accounts for the reduction in the relaxation time.

After 50 days of aging, the bulk structure of the colloidal dispersions has been fairly well established. Separation of the clay platelets and equilibration with the vehicle in the longer chain systems continues beyond 50 days and this causes a reduction in the relaxation time, τ_o, as well as a shift to higher frequencies for tan δ_{max}. The amine-coated clay particles which separate after 50 days can be accommodated by the bulk structure set up in the colloidal dispersion and occupy energy minima which do not lead to increases in ε' and ε''. The increased vehicle-particle interaction thus leads to a drift in τ_o and the frequency of tan δ_{max}, but the ratio of ε'' to ε' (tan δ) does not change appreciably with time accounting for the correlation of tan δ_{max} at 50 days with the sedimentation volumes measured at 100 days of aging. The fact that the relaxation time does not change with shear at a given aging time, as shown in Figure 2, indicates that it does not depend on the degree of agglomeration of the particles but on the interfacial structure and particle size at that time in the aging process.

The time dependency of the dielectric properties of the colloidal dispersion is informative. In the present study, when $\tan \delta_{max}$ reached a constant value, it was indicative of the sedimentation volume as was parameter B at uniform aging times. These trends are related to the interparticle interactions which are present early in the aging process. Sedimentation volumes, however, do not detect the more subtle changes occurring in the systems. The relaxation time, τ_0, is sensitive to the decrease in the particle size of the clay platelets and to the interfacial interactions of the amine chains with the dispersion medium which take place with aging. How these aging processes affect other pertinent physical properties of the colloidal dispersions remains to be investigated.

V. CONCLUSIONS

The interface of a clay-polybutene dispersion is responsible for dielectric relaxations that are not found in either phase taken separately. Dielectric loss amounting to 10% of the dielectric constant was achieved at frequencies around one kilohertz.

The loss maximum increased in magnitude with aging of the dispersion and with concentration of the particles; it decreased with applied shear, and remained independent of temperature. This behavior was attributed to reticulation of the particles via the bipolar layer of adsorbed amine.

The dielectric properties obeyed a skewed circular arc relation characterized by zero- and infinite-frequency dielectric constants, ε'_0 and ε'_∞, relaxation time, τ_0, and skewness parameters, A and B. In general ε'_∞ remained at 2.27, the value for polybutene, whereas the bipolar contribution as represented by ε'_0 increased with increasing concentration, temperature, and chain length of the adsorbed amine.

Reticulation of the particles as represented by low-frequency conductivity was observed at concentrations above 7% (by weight) and at temperatures above 20 C. The resulting structures were destroyed by moderate shear.

Primary effects of the existence and orientation of dipoles were manifested by the relaxation time τ_0 which, while independent of concentration and shear rate, decreased with aging (especially at increased temperatures) because of the gradual perfection of the bipolar layer.

Secondary effects were manifested by the other curve parameters (especially ε'_0), the loss maximum, the conductivity, and the clustering of data in the vicinity of ε'_0. These effects were particularly amenable to study by dielectric relaxation measurements.

Entropic stabilization was evidenced by the smaller sedimentation volumes obtained with the C-6 through C-10 systems, compared with the C-12 through C-16 systems. The longer chain systems experienced primarily steric repulsion which permitted flocculation and resulting higher sediment volumes. An additional contribution from entropic repulsion prevented the short-chain systems from flocculating.

REFERENCES

1. Myers, R. R., Adv. in Chem. 25, 92(1960).
2. Parfitt, G. D., "Dispersion of Powders in Liquids," John Wiley and Sons, New York and Toronto, 1973, p. 273.
3. Lyklema, J., Advan. Colloid & Interface Sci., 2, 65(1968).
4. van der Waarden, M., J. Colloid Sci. 5, 317(1950).
5. Mackor, E. L., J. Colloid Sci. 6, 492(1951).
6. Fischer, E. W., Koll Z., 160, 120(1958).
7. Voet, A., J. Phys. & Colloid Chem., 51, 1037(1947).

8. Bruggeman, D. A. G., Ann. Physik, 24, 636(1935).
9. Dukhin, S. S., "Surface and Colloid Science", E. Matijevic, Ed., Wiley-Interscience, New York, 1971, pp. 83-165.
10. Hanai, T., "Emulsion Science," P. Sherman, Ed., Academic Press, London and New York, 1968, pp. 353-478.
11. Hanai, T., Kolloid-Z., 171, 23(1960).
12. Debye, P., "Polar Molecules", Dover, New York, 1929, p. 94.
13. Cole, K. S., and Cole R. H., J. Chem. Phys., 9, 341(1941).
14. Davidson, D. W., and Cole, R. H., J. Chem. Phys., 19, 1484(1951).
15. Havriliak, S., and Negami, S., J. Plym. Sci., Part C, 14, 99(1966).
16. Oesterle, K. M., and Müller, K. J., J. Paint Tech., 27, 16(1962).
17. Knauss, C. J., Myers, R. R., and Smith P. S., Polymer Letters, 10, 737(1972).
18. Jordan, J. W., J. Phys. Chem. 53, 294(1949).
19. Dawson, P. T. and Haydon, D. A., Kolloid-Z., 203, 133(1965).
20. Shah, M., J. Phys. Chem., 67, 2215(1963).
21. Altshuller, A. P., J. Phys. Chem., 58, 544(1954).

ELECTROPHORETIC BEHAVIOR OF ETHYL CELLULOSE AND POLYSTYRENE MICROCAPSULES CONTAINING AQUEOUS SOLUTIONS OF POLYELECTROLYTES

Akemi Tateno, Motoharu Shiba, and Tamotsu Kondo
Faculty of Pharmaceutical Sciences, Science University of Tokyo, Tokyo, Japan

I. SUMMARY

Ethyl cellulose and polystyrene microcapsules containing aqueous solutions of polyelectrolytes were prepared by an interfacial polymer deposition technique and their electrophoretic mobilities were measured. The microcapsules moved towards the cathode when they contained a cationic polyelectrolyte while they migrated to the anode if an anionic polyelectrolyte was present inside them. Their mobilities decreased as the wall thickness increased. These experimental results were interpreted as showing that polyelectrolyte molecules are adsorbed on the inside surface of the microcapsules and the microcapsule wall has a porous structure that allows counter ions to move along micropores.

II. INTRODUCTION

Previous reports from our laboratory [1, 2] revealed that polyelectrolyte-loaded polyamide microcapsules prepared by an interfacial polymerization technique move towards the anode or cathode in an electric field depending on the sign of electric charge on the encapsulated polyelectrolyte molecules unless they take part in the interfacial polymerization reaction and that the mobility is a function of the polyelectrolyte concentration, and the pH and the ionic strength of the suspending medium. This was interpreted as showing that some of the encapsulated polyions are adsorbed on the inner surface of the microcapsules and a portion of the counter ions of the adsorbed polyions diffuse out of the microcapsules to form an electric double layer around them, the potential of which is obviously affected by the ionic strength of the suspending medium in view of semipermeable nature of the microcapsules [3].

As our knowledge about this type of electric double layer is still scanty, it will be interesting to see, for example, how far the effect of the encapsulated polyions will reach across the membrane. This paper deals with the electrophoretic behavior of polyelectrolyte-loaded microcapsules with much thicker membranes than before as a function of membrane thickness.

III. EXPERIMENTAL

Preparation of microcapsules. As the interfacial polymerization method usually gives microcapsules with ultra thin membranes (10 to 20 nm in thickness) [4, 5], this method of preparation is not suitable to obtain microcapsules of much thicker membranes. Accordingly, an interfacial polymer deposition technique was adopted in the present work and ethyl cellulose and polystyrene microcapsules were prepared by a modification of the procedures described in the patent literatures [6, 7].

Fifty milliliters of aqueous 4% (w/v) gelatin (pI 5.0) solution containing either sodium heparinate (Kokusan Chemicals, Japan) or poly (diallyl dimethyl ammonium chloride) (Calgon Corp., U.S.A.) were mechanically dispersed at 40°C using a constant speed (1,360 r.p.m.) double-bladed stirrer into 50 ml of either ethyl cellulose (Tokyo Kasei Co., Japan) or polystyrene (Mitsubishi-Monsanto Chemical Ind., Japan) solution in dichloromethane to yield a W/O emulsion. The polyelectrolyte concentration in gelatin solution was varied from 0.2 to 1.0% (w/v) while the concentration used of the neutral polymers were 2.5, 5.0, and 10.0% (w/v) in dichloromethane. Stirring was continued for 60 min.

The W/O emulsion thus obtained was then added with stirring to 400 ml of aqueous 1% (w/v) gelatin solution at the same temperature as that of the emulsion to give a W/O/W emulsion. Stirring was again continued for 60 min. The newly formed multiple-phase emulsion was further stirred at room temperature under reduced pressure to allow the organic solvent to evaporate. As the organic solvent evaporated, ethyl cellulose or polystyrene deposited on the surface of droplets of the inner aqueous gelatin solution to form ethyl cellulose or polystyrene microcapsules containing aqueous polyelectrolyte solution suspended in the outer aqueous gelatin solution.

The microcapsules thus formed were repeatedly washed with warm water to remove the gelatin molecules attached to the microcapsule surface and were dialyzed in a Visking tube against distilled water.

Scanning electron micrographs of ethyl cellulose and polystyrene microcapsules are shown in Figs. 1(A) and 1(B), respectively.

Mobility measurement. The electrophoretic mobility measurement of the microcapsules was conducted in a Cytopherometer (Karl Zeiss, West Germany) under a microscope at 25°C. The pH of the medium was kept at 5.0, the isoelectric point of the gelatin used, to eliminate the effect of electric charge of the protein on the mobility while the ionic strength was 0.01. In order to make the measurement at the stationary level of the cell, all necessary precautions were taken. For each measurement, at least 20 microcapsules were timed in each direction to eliminate the polarization effect of the electrodes.

Membrane thickness evaluation. The thickness of the microcapsule membrane was evaluated from the dry weight and the total surface area calculated using the mean diameter of the microcapsules in unit volume of dispersion and the membrane density of the microcapsules determined pycnometrically [5].

IV. RESULTS AND DISCUSSION

As the concentration of the encapsulated polyelectrolytes was increased at a constant membrane thickness the zeta potential increased first and then levelled off. In Figs. 2 and 3 are

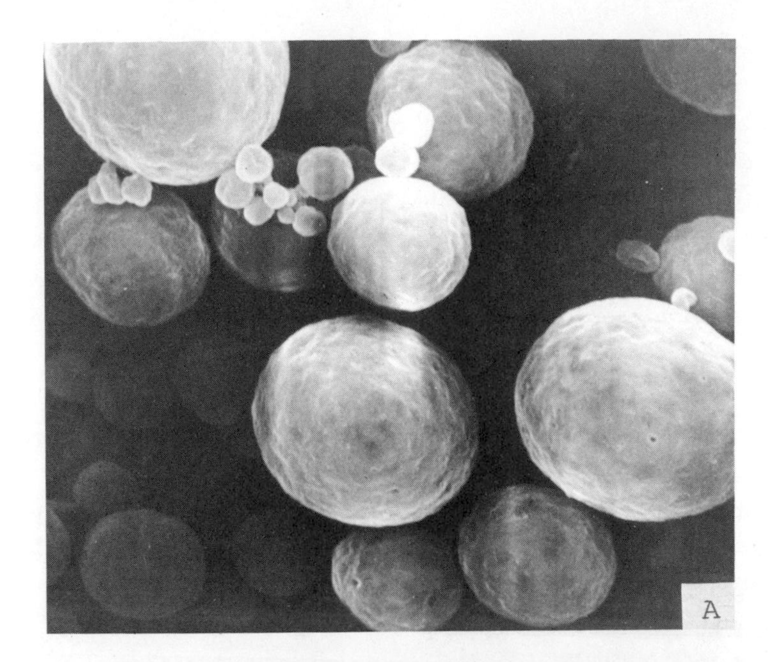

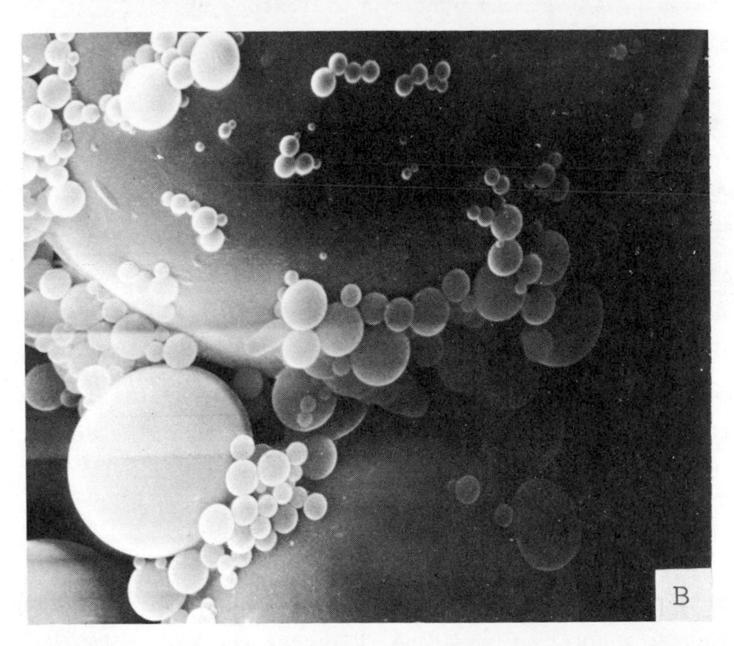

Fig. 1. Scanning electron micrographs of microcapsules prepared by interfacial polymer deposition technique. (A) Ethyl cellulose microcapsules (X 1,000) and polystyrene microcapsules (B) (X 3,000).

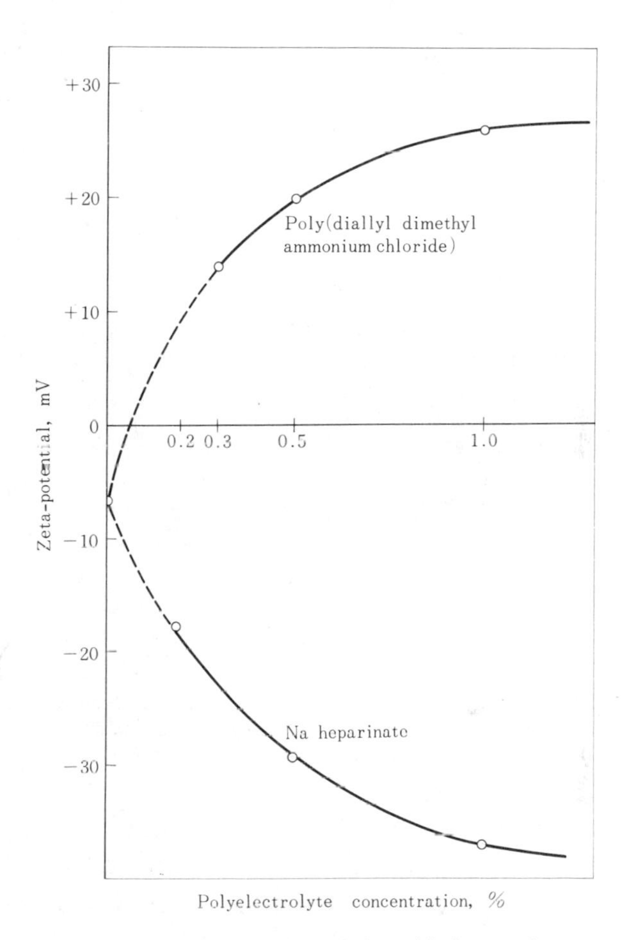

Fig. 2. Zeta potentials of ethyl cellulose microcapsules containing aqueous polyelectrolyte solutions prepared using 2.5% ethyl cellulose solution in dichloromethane as a function of polyelectrolyte concentration at an ionic strength of 0.01.

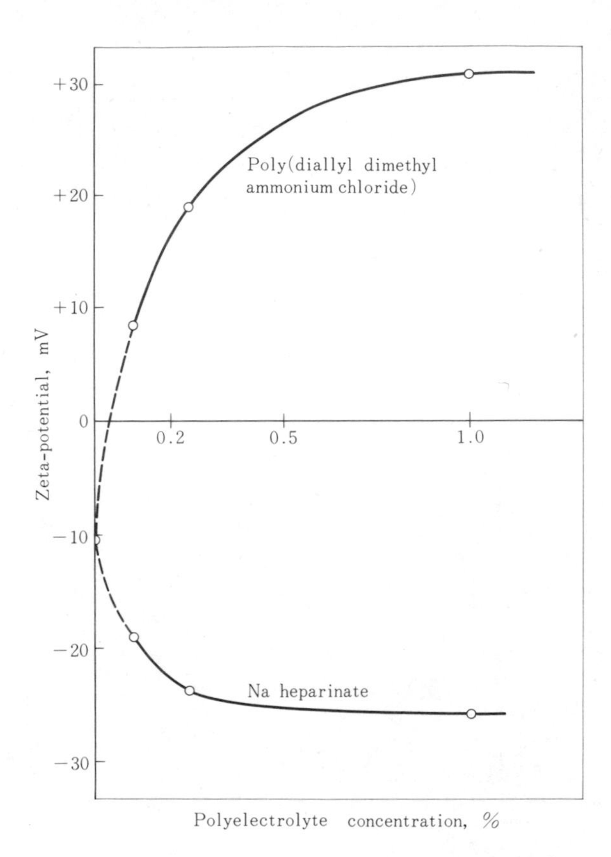

Fig. 3. Zeta potentials of polystyrene microcapsules containing aqueous polyelectrolyte solutions prepared using 2.5% polystyrene solution in dichloromethane as a function of polyelectrolyte concentration at an ionic strength of 0.01.

given the zeta potentials of polyelectrolyte-loaded ethyl cellulose and polystyrene microcapsules as a function of polyelectrolyte concentration, respectively.

The zeta potential clearly reflects the sign of electric charge on the encapsulated polyions. The observed zeta potential-polyelectrolyte concentration relationship suggests a Langmuir type adsorption of the encapsulated polyions on the inner surface of the microcapsules. Since much evidence has been afforded for the applicability of the Langmuir isotherm to the adsorption of serum proteins on polymer membranes [8, 9], it will be reasonable to assume a monolayer adsorption in the present case.

Extrapolation of the curves in the figures to zero polyelectrolyte concentration does not give zero zeta potential but a finite negative value. This is ascribable to a certain number of carboxyl (ethyl cellulose) and sulfate (polystyrene) groups existing on the microcapsule surface as evidenced by the pH dependency of mobility of the microcapsules containing water alone.

Increase in membrane thickness caused a reduction in the zeta potential if the encapsulated polyelectrolyte concentration was kept constant, as shown in Figs. 4 and 5.

As the membrane thickness increases the slipping plane goes farther and farther away from the inner microcapsule surface where a portion of the encapsulated polyions are adsorbed. Then, the zeta potential, the potential at the slipping plane will accordingly decrease.

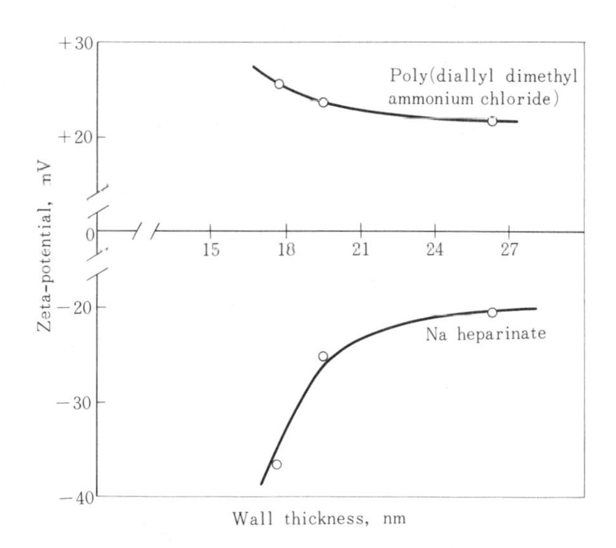

Fig. 4. Zeta potentials of ethyl cellulose microcapsules containing aqueous 1% polyelectrolyte solution as a function of wall thickness at an ionic strength of 0.01.

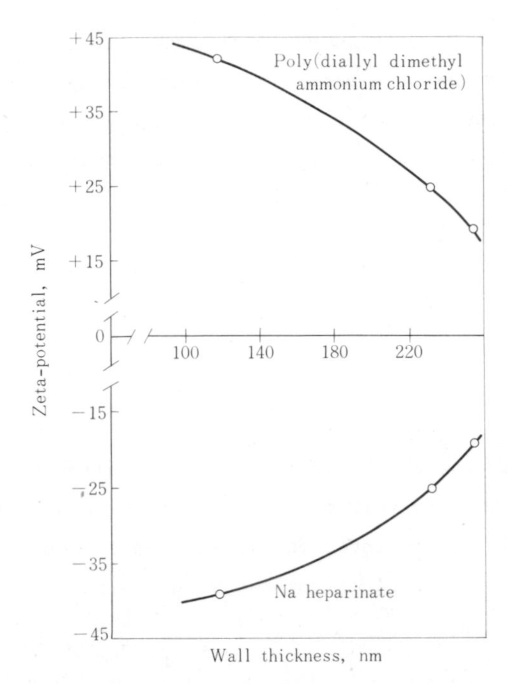

Fig. 5. Zeta potentials of polystyrene microcapsules containing aqueous 1% polyelectrolyte solution as a function of wall thickness at an ionic strength of 0.01.

Since the microcapsule membrane is quite likely to have a porous structure, potential change with the distance from the inner microcapsule surface across the membrane will be given by the following equation:

$$y_1 = 4\gamma e^{-\sqrt{1-\alpha}\kappa x_1} \qquad (1)$$

where y_1 is $ze\psi_1/kT$ in which ψ_1 is the potential at a distance x_1 from the inner surface and is assumed to be equal to the experimentally determined zeta potential, γ is $\tanh ze\psi_0/4kT$ in which ψ_0 is the potential on the inner microcapsule surface, and α is the fraction of the total space within the membrane which is not available to counter ions [10, 11]. In the outside region of the membrane, potential will vary in accordance with the equation:

$$y_2 = 4\gamma e^{-\kappa x_2} . \qquad (2)$$

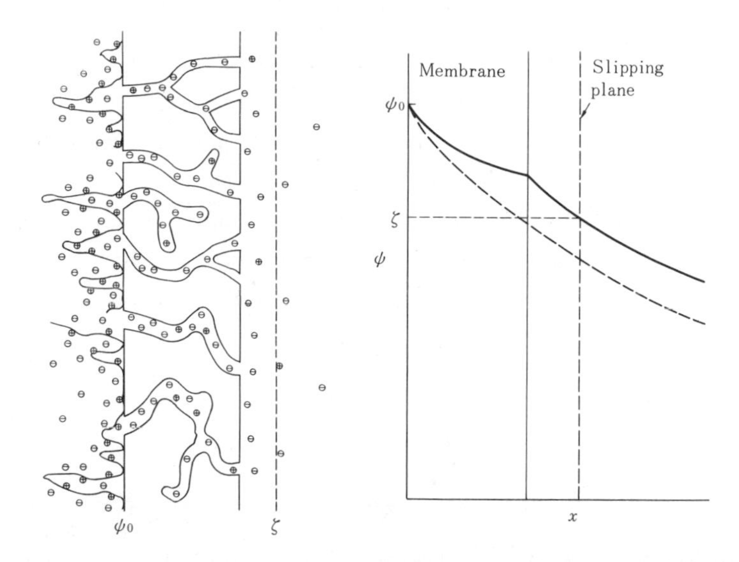

Fig. 6. Model of electric double layer around the inner microcapsule surface and schematic representation of potential change based on the model.

Furthermore, in the hypothetical case where no polymer membrane covers the adsorbed layer of polyions, that is, when $\alpha = 0$, potential drop with increasing distance from the adsorbed polyion layer will be given by

$$y_1' = 4\gamma e^{-\kappa x_1} \quad . \tag{3}$$

This situation is schematically shown in Fig. 6. As α lies between 1 and 0, y_1' drops more rapidly than does y_1 when x_1 is increased.

It is now possible for us to evaluate the value of α using the data in Fig. 4 for ethyl cellulose microcapsules containing aqueous polyelectrolyte solutions in the following way. Extrapolation of the curves in the figure to zero wall thickness gives 50 mV and -75 mV as the values of ψ_0 of the microcapsules containing the cationic and anionic polyelectrolytes, respectively. Putting $\kappa = 0.331\ nm^{-1}$ and $\psi_0 = 50$ mV or -75 mV in Eq. 3, y_1' can be calculated as a function of x_1. Finally, α is obtained from the ratio, y_1/y_1'. The value of α as evaluated in this way is found to lie between 0.97 and 0.99. This means that only 1 to 3% of the total space within ethyl cellulose membrane is available to counter ions.

No such calculation was carried out on polystyrene microcapsules because it was feared that erroneous ψ_0 values may result from very long extrapolation of the curves in Fig. 5.

Acknowledgement

The authors thank Messrs. A. Chayama and K. Kajino for their experimental assistance in this work.

References

1. Shiba, M., Kawano, Y., Tomioka, S., Koishi, M., and Kondo, T., Kolloid Z. Z. Polymere, 249, 1056 (1971).

2. Shiba, M., Kawano, Y., Tomioka, S., Koishi, M., and Kondo, T., Bull. Chem. Soc. Japan, 44, 2911 (1971).

3. Takamura, K., Koishi, M., and Kondo, T., Kolloid Z. Z. Polymere, 248, 929 (1971).

4. Chang, T. M. S. and Poznansky, M. J., J. Biomed. Mater, Res., 2, 187 (1968).

5. Takamura, K., Koishi, M., and Kondo, T., J. Pharm. Sci., 62, 610 (1973).

6. French Patent 1,362,933 (1964).

7. Japanese Patent S42-13703 (1967).

8. Brash, J. L. and Lyman, D. J. in "The Chemistry of Biosurfaces", Vol. 1 (Hair, M. L., ed.) p.199. Marcel Dekker, New York, 1971.

9. Dillman, W. J. and Miller, I. F., J. Colloid Interface Sci., 44, 221 (1973).

10. Haydon, D. A., Biochim. Biophys. Acta, 50, 450 (1961).

11. Haydon, D. A. and Taylor, F. H., Phil. Trans. Roy. Soc. London, A252, 225 (1960).

THE DORN EFFECT IN A CONCENTRATED SUSPENSION OF SPHERICAL PARTICLES

Samuel Levine[†]

Chemical Engineering Department
University of British Columbia
Vancouver, Canada

Graham Neale

Chemical Engineering Department
University of Ottawa
Ottawa, Canada

Norman Epstein

Chemical Engineering Department
University of British Columbia
Vancouver, Canada

When a swarm of charged particles settles under gravity, a vertical electric field E is induced which acts upon the particle charge to produce a retardation in settling velocity relative to that which would be observed if the particles were uncharged. In this work a geometric cell model is employed to study these effects. When the particles are sedimenting the potential distribution around each particle is perturbed away from its spherically symmetric equilibrium (i.e. no particle motion) distribution by an amount proportional to $[Xr + mr^{-2} + g(r)] \cos\theta$, where the Xr term accounts for the perturbation due to the induced electric field, the mr^{-2} term accounts for that due to the dipole moment, m, associated with each particle, and g(r) denotes a short-range function which accounts for the distortion of the charge distribution in the diffuse layer due to the particle motion. In a previous work the rather over-simplified expression $g(r) = Pe^{-\kappa r}$ was adopted for the short-range function. The purpose of the present work is to derive a rigorous expression for g(r) for use in predicting the induced sedimentation field and the retardation in settling velocity. The analysis is confined to the case of 1-1 electrolytes having equal mobilities for cations and anions and is restricted to small values of surface potential (in the Debye-Huckel range).

I. INTRODUCTION

It is well known that when a suspension of charged particles settles under gravity through an electrolyte medium, a vertical

† Permanent Address: Mathematics Department, University of Manchester, Manchester, England.

electric field E is induced. This phenomenon is frequently referred to as the Dorn Effect. The induced electric field exerts a retarding force on the particles. In practice, the observed settling velocity is often considerably less than that which would prevail if the particles were uncharged (1). The first theoretical estimates of the above phenomena were given by Smoluchowski (2). He deduced that for a very dilute suspension of identical non-conducting spheres of radius a, in which the double layer thickness $1/\kappa$ was very much less than a (i.e. $\kappa a >> 1$),

$$E = \varepsilon \, \psi_a \, (\rho_s - \rho_l) \, cg/4\pi \, \lambda \, \mu \qquad [1.1]$$

where ψ_a denotes the surface-potential, ε, λ and μ the dielectric constant, specific conductivity and viscosity of the neutral electrolyte respectively, ρ_s and ρ_l the solid and liquid densities, g the acceleration due to gravity, and c the fractional volumetric concentration of particles. For a single isolated spherical particle he showed that (for $\kappa a >> 1$)

$$\frac{U_o}{U} = 1 + \frac{\varepsilon^2 \, \psi_a^2}{16\pi^2 a^2 \lambda \, \mu} \qquad [1.2]$$

where U is the actual settling velocity and U_o that which would be observed if the same particle was uncharged. Booth (3) generalized [1.2] to arbitrary values of κa and to the case of very dilute suspensions of spheres ($c \simeq 0$). Sengupta (4) subsequently presented some useful numerical estimates based on Booth's rather complicated theory. More recently, Levine et al (1) extended [1.1] and [1.2] in an approximate manner to arbitrary values of κa for the practically important case of a concentrated suspension of spheres (i.e. $0 \leqslant c \lesssim 0.5$). The effects of particle - particle interactions, which were implicitly neglected by Smoluchowski and Booth, were accounted for by employing a familiar cell model technique which had previously provided good predictions for the related problem of the electrophoresis of a concentrated suspension of spheres (5), (6). The predictions obtained for sedimentation (1) were in qualitative agreement with the rather sparse experimental data available in the literature.

In the cell model, each sedimenting particle (of radius a) is envisaged as being surrounded by a diffuse layer of ions which takes the form of a hypothetical concentric shell of electrolyte [refer to Fig. 1 in Reference (1)] containing just sufficient ions to neutralize the particle charge, and possessing an outer radius b such that all cells taken together just fill the total suspension volume [i.e. $c = (a/b)^3$]. For mathematical convenience, we consider the particles to be stationary and the fluid to be streaming upwards with a uniform mean velocity U.

In the (hypothetical) equilibrium situation, when no motion is occurring, the potential distribution $\psi(r)$ within the cell is spherically symmetric and is given by [2.4]. However, when the particles are undergoing sedimentation the potential is perturbed away from spherical symmetry, the perturbation in potential being given by [2.17]. In this equation, the term involving X accounts for the potential due to the globally induced sedimentation field (Dorn Potential), m denotes the dipole moment arising from the charge displacement in the diffuse layer and the g(r) term is a "short-range" contribution which results from the loss of spherical symmetry of the diffuse layer and which corresponds to the relaxation term in electrophoresis. In the earlier work (1), the crude but simple functional form $g(r) = Pe^{-\kappa r}$, where P is a constant, was assumed. The principal contribution of the present work is to describe how a more rigorous expression $g(r) = P(r)e^{-\kappa r} + Q(r)e^{\kappa r}$ may be derived for the case of small zeta-potentials ψ_a.

As shown previously (1), the components of the fluid velocity vector $\underline{u}$ for incompressible fluid flow within the cell are:

$$u_r = U.\ w(r).\ \cos\theta \qquad [1.3]$$

$$u_\theta = U.\ \hat{w}(r).\ \sin\theta \qquad [1.4]$$

where

$$w(r) = -(A_1/2)(r/a)^2 - A_2 - 2A_3(a/r) + 2A_4(a/r)^3 \qquad [1.5]$$

$$\hat{w}(r) = A_1(r/a)^2 + A_2 + A_3(a/r) + A_4(a/r)^3 \qquad [1.6]$$

with A_1, A_2, A_3 and A_4 denoting arbitrary constants which depend upon the particular choice of hydrodynamic boundary conditions at r = b. For the Kuwabara model, employed previously (1), these constants possess the values:

$$A_1 = -3c/J,\ A_2 = 5(2 + c)/2J,\ A_3 = -15/4J,$$

$$A_4 = (2c-5)/4J,\ J = 5-9c^{1/3} + 5c-c^2,\ c = (a/b)^3 \qquad [1.7]$$

It should be noted that we are here neglecting any contribution to $\underline{u}$ due to electrical effects. That is, we are neglecting any small changes in the fluid flow field which occur as a result of the motion of the ions in the induced electric field (this corresponds to the electrophoretic retardation effect in electrophoresis). This effect is proportional to ψ_a^2 and hence is of a second order nature at small ψ_a (1). Thus, in the subsequent analysis, [1.3]-[1.7] will be adopted to describe the bulk fluid flow and terms of order ψ_a^2 and higher appearing in the following treatment will be neglected. The overall analysis will therefore be valid to order ψ_a.

II. ELECTRICAL CONSIDERATIONS

At equilibrium, in the absence of any particle motion, the number densities $n_{\pm}^{o}$ of the two ion species are given by the Boltzmann equation. For a 1-1 electrolyte, to which we shall confine our attention, the densities are assumed to be:

$$n_{\pm}^{o} = n_{\pm}^{o}(r) = n \cdot \exp[\mp e_o \psi(r)/kT] \qquad [2.1]$$

where $\psi(r)$ is the spherically symmetric potential at distance r from the center of the reference particle, n the number density of cations or anions at some reference point where the electrostatic potential is assigned to be zero (6), e_o is the proton charge, k Boltzmann's constant and T the absolute temperature. For small enough potentials we need retain only linear terms in $\psi(r)$, as discussed previously, and upon substituting [2.1] into Poisson's equation we obtain the linearized Poisson - Boltzmann equation

$$\nabla^2 \psi(r) = \kappa^2 \psi(r) \qquad [2.2]$$

where κ denotes the familiar Debye - Huckel parameter, given by

$$\kappa^2 = 8\pi n e_o^2/\varepsilon kT \qquad [2.3]$$

If the suspension of particles completely and uniformly fills the volume of the solution then in [2.2] - [2.3] we should really identify n with $\sqrt{n_+ n_-}$ (which remains constant throughout the diffuse layer in the cell) and the potential ψ with that measured relative to a potential given by $\psi_r = (kT/2e_o)\ln(\bar{n}_+/\bar{n}_-)$, where $\bar{n}_+$, $\bar{n}_-$ denote the densities of cations and anions at some convenient common reference point within the cell (6). The most convenient (but not necessarily the best) choice of reference potential would be $\psi(b)$. The use of [2.2] in the cell model therefore implies a small potential difference, $\psi(b) - \psi(a)$, rather than a small absolute value of $\psi(a)$.

The solution of [2.2] subject to the stipulated boundary conditions $\psi(a) = \psi_a$ and $(d\psi/dr)_{r=b} = 0$ is (1):

$$\psi(r) = (a/r)(N_1 e^{-\kappa r} + N_2 e^{\kappa r})\psi_a \qquad [2.4]$$

where

$$N_1 = e^{\kappa a}/(1+L), \; N_2 = Le^{-\kappa a}/(1+L), \; L = \left[\frac{\kappa b+1}{\kappa b-1}\right] e^{2\kappa(a-b)} \qquad [2.5]$$

Equation [2.4] is valid to order ψ_a^2.

When the particles are undergoing sedimentation, the densities of the two ion species become (3):

$$n_{\pm} = n_{\pm}^{o} + m_{\pm} \qquad [2.6]$$

where m_+, m_- denote the "perturbation" densities due to the motion of the particles. Similarly, the diffuse layer around each particle is altered and the potential distribution within each cell now becomes $\psi + \phi$, where ϕ denotes the perturbation potential. The velocity vectors of the ions relative to the particle center are (3):

$$\underline{u}_{\pm} = \underline{u} - (kT/f_{\pm})\ \text{grad}\ (\ln n_{\pm}) \mp (e_o/f_{\pm})\ \text{grad}\ (\psi + \phi) \qquad [2.7]$$

where $\underline{u}$ is the velocity of the solvent, and $f_{\pm}$ the friction coefficients of the ion species. Assuming steady state flow, the continuity requirement for each ion species is that

$$\text{div}\ (n_{\pm}\ \underline{u}_{\pm}) = 0 \qquad [2.8]$$

Substituting [2.6] and [2.7] into [2.8], and making use of [2.1] in its equivalent form e_o grad $\psi(r) = \mp kT$ grad$[\ln n_{\pm}^{o}(r)]$, finally leads to

$$\mp e_o \text{div}(n_{\pm}\text{grad}\phi + m_{\pm}\text{grad}\psi) - kT\nabla^2 m_{\pm} + f_{\pm}\text{div}(n_{\pm}\underline{u}) = 0 \qquad [2.9]$$

Provided the fluctuation or self-atmosphere potential terms are neglected in the Boltzmann relations for the ion densities, [2.1], [2.7], [2.8] and consequently [2.9] represent mathematically exact relationships.

Following Booth (3), we neglect terms of order U^2 whence products of $m_{\pm}$ with $m_{\pm}$, ϕ and components of $\underline{u}$ may be neglected in [2.9]. Furthermore, ψ and $n_{\pm}^{o}$ are functions only of r and, assuming the solvent to be incompressible, div $\underline{u} = 0$. It is now readily verified that [2.9] reduces to

$$e_o[n_{\pm}^{o}\nabla^2\phi + m_{\pm}\nabla^2\psi + (\partial\phi/\partial r)(dn_{\pm}^{o}/dr) + (\partial m_{\pm}/\partial r)(d\psi/dr)]$$

$$\pm kT\nabla^2 m_{\pm} \mp f_{\pm}u_r(dn_{\pm}^{o}/dr) = 0 \qquad [2.10]$$

where u_r is given by [1.3]. Equation [2.10] is equivalent to Equation [2.6] in Booth's work (3). For simplicity in the forthcoming analysis we will assume that $f_+ = f_- = f$.

We now proceed rather differently from Booth, by introducing

$$M = m_+ - m_-, \quad M^* = m_+ + m_- \qquad [2.11]$$

whence Poisson's equations, applied in the equilibrium and perturbed states respectively, namely

$$\nabla^2\psi = -4\pi e_o(n_+^o - n_-^o)/\varepsilon \ , \quad \nabla^2(\psi + \phi) = -4\pi e_o(n_+ - n_-)/\varepsilon \qquad [2.12]$$

can be subtracted to yield

$$\nabla^2\phi = -4\pi e_o M/\varepsilon \qquad [2.13]$$

Also, expanding [2.1] in powers of ψ yields

$$n_+^o + n_-^o = 2n[1 + (e_o\psi/kT)^2/2 + . . .]$$

$$n_+^o - n_-^o = - (2ne_o\psi/kT)[1 + (e_o\psi/kT)^2/6 + . . .] \qquad [2.14]$$

Then, upon adding and subtracting the two equations in [2.10] and making use of [2.2], [2.11], [2.13] and [2.14], we finally obtain

$$\nabla^2 M-\kappa^2 M = -e_o[\kappa^2\psi M^*+(\partial M^*/\partial r)(d\psi/dr)+(2fnu_r/kT)(d\psi/dr)]/kT \qquad [2.15]$$

$$\nabla^2 M^* = -e_o[2\kappa^2\psi M+(\partial M/\partial r)(d\psi/dr)-(2ne_o/kT)(\partial\phi/\partial r)(d\psi/dr)]/kT \qquad [2.16]$$

which are valid to order ψ_a.

With the cell model, the dependence of $\phi = \phi(r,\theta)$ and $m_\pm = m_\pm(r,\theta)$ on angle θ is described by the factor $\cos\theta$. To a first approximation, the departure from the spherical symmetry of the potential distribution around a stationary particle can be expressed in the form (1)

$$\phi(r,\theta) = U[Xr + mr^{-2} + g(r)]\cos\theta \qquad [2.17]$$

The first two terms, characterized by the parameters X and m (which need to be determined by application of the appropriate boundary conditions) denote the "long-range" contributions to $\phi(r,\theta)$ whereas g(r) is a "short-range" term (8), which corresponds to the relaxation term when studying the electrophoresis of a particle under an externally applied electric field. In this work we shall show how to derive an expression for g(r) which is more rigorous than the simple form used previously (1).

Introducing

$$M = U.v(r).\cos\theta \ , \ M^* = U.v^*(r).\cos\theta \qquad [2.18]$$

and making use of [1.3], [2.17] and [2.18], the angle θ may be eliminated from [2.15] and [2.16] to yield two ordinary differential equations in the variable r :

$$d^2v/dr^2 + (2/r)(dv/dr) - (\kappa^2r^2 + 2)v/r^2 = G(r) \qquad [2.19]$$

$$d^2v^*/dr^2 + (2/r)(dv^*/dr) - 2v^*/r^2 = G^*(r) \qquad [2.20]$$

where, after deleting terms of order $\psi_a{}^2$ and higher,

$$G(r) = -e_o[(2fnw/kT)(d\psi/dr)]/kT \qquad [2.21]$$

$$G^*(r) = -e_o[2\kappa^2\psi v + (d\psi/dr)(dv/dr) - (2ne_o/kT)(X - 2mr^{-3} + \{dg/dr\})(d\psi/dr)]/kT \qquad [2.22]$$

which are valid to order ψ_a. Provided $f_+ = f_-$, the function $G^*(r)$ is of order $\psi_a{}^2$.

General solutions of [2.19] and [2.20] are

$$v(r) = Ae^{-\kappa r}(\kappa r+1)/\kappa r^2 + Be^{\kappa r}(\kappa r-1)/\kappa r^2 + \int_a^r G(z).h(r,z).dz \qquad [2.23]$$

$$v^*(r) = Cr + Dr^{-2} + (1/3)\int_a^r G^*(z).(r - z^3r^{-2}).dz \qquad [2.24]$$

where

$$h(r,z) = \frac{z}{2\kappa r}\left[\frac{(\kappa z+1)(\kappa r-1)}{\kappa^2 zr}e^{-\kappa(z-r)} - \frac{(\kappa z-1)(\kappa r+1)}{\kappa^2 zr}e^{\kappa(z-r)}\right] \qquad [2.25]$$

the particular integrals in [2.23] and [2.24] being obtained by the variation of parameters method. The integrations involved in [2.23] and [2.24] are essentially elementary but excessively tedious.

Calculation of $g(r)$

By making use of [2.17] and [2.18], and noting that the first two terms in [2.17] satisfy Laplace's equation, we can eliminate the angle θ from [2.13] to obtain

$$d^2g/dr^2 + (2/r)(dg/dr) - (2g/r^2) = -4\pi e_o v/\varepsilon \quad [2.26]$$

A particular integral of [2.26] is

$$g(r) = -(4\pi e_o/3\varepsilon)\int_a^r (r - z^3r^{-2}).v(z).dz \quad [2.27]$$

again using the variation of parameters method. (The general solution of the corresponding homogeneous part of [2.26] is $P_1r + P_2/r^2$, which may therefore be absorbed into the first two terms of [2.17]). We therefore have an expression for $g(r)$, namely [2.27], which contains, via the function v, the two arbitrary constants A and B, and which is correct to order ψ_a. We anticipate that, after simplification, [2.27] will reduce to the form $g(r) = P(r)e^{-\kappa r} + Q(r)e^{\kappa r}$ where $P(r)$ and $Q(r)$ are slowly varying functions of r.

In all, therefore, we have six unknown arbitrary constants, A, B, C, D, X and m, so consequently require six independent physically realistic boundary conditions to permit their determination.

III. BOUNDARY CONDITIONS

Boundary Conditions at $r = a$

Since there is no free charge <u>inside</u> the particles, the interior potential, ψ_i, is governed by Laplace's equation $\nabla^2\psi_i = 0$, having a solution of the general form $\psi_i = A + Br\cos\theta$. Applying Gauss' theorem in both the equilibrium (stationary) and perturbed states yields the conditions (8)

$$4\pi\sigma_o = -\varepsilon(d\psi/dr) = -\varepsilon[\partial(\psi+\phi)/\partial r] + \varepsilon_i(\partial\psi_i/\partial r) \quad \text{at } r=a \quad [3.1]$$

where ε_i denotes the dielectric constant of the particle and σ_o the uniform (fixed) charge density on the particle surface. Continuity of potential requires that $\psi_i = \psi + \phi$ at $r = a$. Applying this condition in conjunction with [3.1] and [2.17], and noting from [2.27] that $g(a) = g'(a) = 0$, finally yields the relation

$$X(\varepsilon - \varepsilon_i) = m(2\varepsilon + \varepsilon_i)/a^3 \quad [3.2]$$

Since the spheres are impermeable the radial velocity components of the two ion species, in addition to that of the solvent molecules, relative to the particle center must vanish at r = a, namely

$$u_{r-} = u_{r+} = 0 \quad \text{at} \quad r = a \qquad [3.3]$$

Incorporating [3.3] with [2.7] and [2.6], adding and subtracting the resulting equations, then applying [2.11] and [2.14] yields the boundary conditions

$$\left[(\partial M^*/\partial r) + (e_o M/kT)(d\psi/dr) - (2ne_o{}^2/k^2T^2)\psi(\partial\phi/\partial r)\right]_{r=a} = 0 \qquad [3.4]$$

$$\left[(\partial M/\partial r) + (e_o M^*/kT)(d\psi/dr) + (2ne_o/kT)(\partial\phi/\partial r)\right]_{r=a} = 0 \qquad [3.5]$$

Equations [3.4] and [3.5] are of orders $\psi_a{}^2$ and ψ_a respectively. Both are required in order to provide sufficient conditions for determining the six constants introduced above. Making use of [2.17] and [2.18], [3.4] becomes

$$\left[(dv^*/dr) + (e_o v/kT)(d\psi/dr)\right]_{r=a} = 2ne_o{}^2\psi_a(X - 2ma^{-3})/k^2T^2 \qquad [3.6]$$

and, to order ψ_a^2, [3.5] simplifies to

$$(dv/dr)_{r=a} = - 2ne_o \ (X - 2ma^{-3})/kT \qquad [3.7]$$

In the steady state, it is necessary (1), (3) that the net flow of electric current across any horizontal plane due to fluid convection must be equal and opposite to that due to electric conduction under the influence of the induced electric field. In the modelled system, this condition is most easy to formulate at the plane $\theta = \pi/2$, thus:

$$2\pi \int_a^b r.e_o.(u_{\theta+}n_+ - u_{\theta-}n_-).dr + 2\pi \int_a^b r.i_\theta.dr = 0 \quad \text{at } \theta = \pi/2 \qquad [3.8]$$

where

$$i_\theta(r,\theta) = - (\lambda/r)(\partial\phi/\partial\theta) \qquad [3.9]$$

denotes the tangential component of the current density vector, and λ is the conductivity of the electrolyte within the cell. Neglecting surface conductance effects (7), λ may be identified with the conductivity of the neutral electrolyte, so $\lambda = 2ne_o^2/f$ for a 1-1 electrolyte having equal friction coefficients (mobilities) for cations and anions (8). Combining [3.8] and [3.9] with [1.4], [2.6], [2.7], [2.11], [2.17] and [2.18] yields, to order ψ_a,

$$e_o \int_a^b [2n\hat{w}re_o f\psi - k^2T^2 v - 2ne_o kT(Xr + mr^{-2} + g)]\, dr$$

$$= fkT\lambda \int_a^b (Xr + mr^{-2} + g)\, dr \qquad [3.10]$$

Recalling that $\lambda = 2ne_o^2/f$, [3.10] simplifies to

$$\int_a^b (2n\hat{w}re_o{}^2 f\psi - k^2T^2 e_o v)\, dr = 2fkT\lambda \int_a^b (Xr + mr^{-2} + g)\, dr \qquad [3.11]$$

Boundary Conditions at r = b

At the outer envelope (r = b) of the cell, we assume the hydrodynamic boundary condition used by Happel and by Kuwabara (10) for the radial component of the solvent velocity

$$u_r = - U \cos\theta \qquad [3.12]$$

Also, the radial component of the local induced electric field is equated to the radial component of the exterior macroscopically uniform induced electric field E, that is

$$- [\partial(\psi + \phi)/\partial r]_{r=b} = - [\partial\phi/\partial r]_{r=b} = - E \cos\theta \qquad [3.13]$$

recalling that $(\partial\psi/\partial r)_{r=b} = 0$ from [2.4]. The second equation in [3.13] provides the relation between the induced electric field E and the actual velocity of settling U

$$E = U[X - 2mb^{-3} + g'(b)] \qquad [3.14]$$

after invoking [2.17]. Both E and U, the two most important macroscopic quantities associated with the Dorn Effect, are easily measurable in the laboratory. It follows from [2.7], [3.12] and [3.13] that the radial components of the ion velocities at r = b are

$$(u_{\pm})_r = -U\cos\theta \mp (e_o/f)E\cos\theta - (kT/f)[(1/n_{\pm})(dn_{\pm}/dr)]_{r=b} \qquad [3.15]$$

When $b \to \infty$, this must reduce to

$$(u_{\pm})_r = - U \cos\theta \mp (e_o/f)E\cos\theta \qquad [3.16]$$

which Booth (3) employed for a single isolated particle, suggesting the boundary conditions

$$dn_{\pm}/dr = 0 \qquad \text{at } r = b \qquad [3.17]$$

The physical interpretation of [3.17] is that there is no ion transport across the outer envelope r = b due to diffusional forces. In other words, any macroscopic ion transport between adjacent cells occurs only as a result of hydrodynamic and electrical forces. This seems very plausible physically on a locally averaged basis since there is no macroscopic gradient of $n_{\pm}$ in the real system. Combining [3.17] with [2.6] yields

$$(\partial n^o_{\pm}/\partial r) + (\partial m_{\pm}/\partial r) = 0 \qquad \text{at } r = b \qquad [3.18]$$

However, it follows from [2.1] and the condition $(d\psi/dr)_{r = b} = 0$ that the first term in [3.18] is identically zero, whence this equation reduces to

$$\partial m_{\pm}/\partial r = 0 \qquad \text{at } r = b \qquad [3.19]$$

Adding and subtracting the two equations in [3.19], in conjunction with [2.11] and [2.18], finally yields the sought pair of boundary conditions

$$dv(r)/dr = 0 , \quad dv^*(r)/dr = 0 \qquad \text{at } r = b \qquad [3.20]$$

Equations [3.2], [3.6], [3.7], [3.11] and [3.20] represent six independent relations involving, via v and v*, the six unknown arbitrary constants A, B, C, D, X and m. Their simultaneous solution is essentially straightforward, yielding expressions in terms of the dimensionless parameters κa and c. However, the derivation of these expressions is excessively time consuming. Work is currently proceeding on their evaluation.

IV. RESULTS AND DISCUSSION

After substituting for the arbitrary constants X, m, A and B in terms of κa and c, [3.14] can be expressed in the more general form obtained earlier (1), namely

$$E = (9\varepsilon\psi_a Uc/8\pi\lambda a^2)\ \gamma(\kappa a,c) \qquad [4.1]$$

where $\gamma(\kappa a,c)$ denotes a very complicated function of κa and c. Previously (1), an analytical expression for $\gamma(\kappa a,c)$ was obtained based upon the crude but simple form $g(r) = Pe^{-\kappa r}$. The predictions obtained therefrom are re-discussed here since those based upon the more rigorous present theory are not yet available. It will be recalled that for large κa, $\gamma(\kappa a,c) \to 1$ as $c \to 0$. Under these conditions we have a system in which hydrodynamic interaction between particles is negligible whence $U \simeq U_o = 2a^2(\rho_s - \rho_l)g/9\mu$ (from Stokes Law for a single particle) and [4.1] reduces to Smoluchowski's result [1.1] as required. As c increases γ increases appreciably; however, E does not increase in proportion since U decreases as c increases. Hence, to obtain an explicit prediction for E we need to develop an expression for U in terms of U_o (where U_o denotes the settling velocity which would be observed if the spheres were uncharged).

Calculation of U

In the steady state, the net force acting on any typical sphere is zero, that is

$$F_h + F_e - (4/3)\pi a^3(\rho_s - \rho_l)g = 0 \qquad [4.2]$$

where F_h denotes the vertical component of the hydrodynamic force acting on the sphere (due to the motion of the fluid past it), F_e denotes the vertical component of the electrical force acting on the sphere (due to the various interactions between the local induced electric field and the charges on the particle and within the diffuse layer) and the third term denotes the force of gravity acting on the sphere. Based on the Kuwabara model (1)

$$F_h = 6\pi\mu a U\Omega \qquad [4.3]$$

where

$$\Omega = 5/(5 - 9c^{1/3} + 5c - c^2) \qquad [4.4]$$

It should be noted that [4.3] is based on [1.3] and [1.4], in which the electro-osmotic effects of the ion movement on the solvent flow field (and hence on the viscous drag force exerted on the particle) have been neglected. This second-order[1] effect corresponds to the electrophoretic retardation effect in electrophoresis.

The electrical force on the particle consists of two components. The first is the force of the local induced field on the particle charge. The second is the retardation force which arises because the center of the diffuse layer charge lags behind the center of the particle charge (this effect is analagous to the relaxation effect in electrophoresis). The total electrical force on the particle in the vertical z-direction is given by Wiersema et al (9):

$$F_e = - \int_S \sigma(\mathrm{grad}\Psi \bullet \underline{k})\, dS \qquad [4.5]$$

where $\Psi = \psi(r) + \phi(r,\theta)$, σ denotes the surface charge density on the particle (which we assume to be uniform and constant), $\underline{k}$ denotes the unit vector in the z-direction, and S the surface of the sphere. Simplifying [4.5] and noting that $dS = 2\pi a^2 \sin\theta d\theta$ yields:

$$F_e = - 2\pi a^2\sigma \int_0^\pi [\frac{\partial\phi}{\partial z} - \frac{d\psi}{dr}\cos\theta]_{r=a} \sin\theta\, d\theta \qquad [4.6]$$

$$= - 2\pi a^2\sigma \int_0^\pi (\partial\phi/\partial z)_{r=a} \sin\theta\, d\theta \qquad [4.7]$$

the $d\psi/dr$ term in [4.6] making no contribution since $\int_0^\pi \cos\theta\sin\theta d\theta = 0$. Resolution of vectors in the z-direction yields

$$(\partial\psi/\partial z) = (1/r)(\partial\phi/\partial\theta)\sin\theta - (\partial\phi/\partial r)\cos\theta \qquad [4.8]$$

Combining [4.7], [4.8] and [2.17], and recalling that $g(a) - g'(a) = 0$, finally yields the simple expression

$$F_e = 4\pi a^2\sigma UX \qquad [4.9]$$

with $4\pi a^2\sigma$ representing the total charge on the particle surface. From Gauss' theorem $\sigma = - (\varepsilon/4\pi)(d\psi/dr)_{r=a}$ whence

$$F_e = - \varepsilon a^2 UX(d\psi/dr)_{r=a} \qquad [4.10]$$

Combining [4.2], [4.3] and [4.10] yields

$$6\pi\mu aU\Omega - \varepsilon a^2 UX(d\psi/dr)_{r=a} = 4\pi a^3(\rho_s - \rho_l)g/3 \qquad [4.11]$$

The equation corresponding to [4.11] for the case of an uncharged swarm reads

$$6\pi\mu aU_o\Omega = 4\pi a^3(\rho_s - \rho_l)g/3 \qquad [4.12]$$

Dividing [4.12] by [4.11] yields an expression for U_0/U, thus

$$U_o/U = 1 - (\varepsilon a^2 X/6\pi\mu a\Omega)(d\psi/dr)_{r=a} \qquad [4.13]$$

After substituting for the arbitrary constant X in terms of κa and c, and for $d\psi/dr$ from [2.4], [4.13] can be expressed in the more general form obtained earlier (1)

$$U_o/U = 1 + (\varepsilon^2\psi_a^2/16\pi^2 a^2\lambda\mu)H(\kappa a,c) \qquad [4.14]$$

where $H(\kappa a,c)$ denotes a very complicated function of κa and c. (This expression can then be introduced into [4.1] thereby producing a direct prediction for E, the overall induced electric field). The velocity, U_0, of an uncharged swarm can be calculated by a number of methods (10), and expressed in terms of the Stokes velocity $U_0^* = 2a^2 (\rho_s - \rho_l)\, g/9\mu$ of a single particle in an infinite fluid medium. Thus, the currently employed Kuwabara model predicts, via [4.12], that $U_0/U_0^* = 1/\Omega$. Since predictions based on the present theory are not yet available, we again refer to those based upon the simple form $g(r) = Pe^{-\kappa r}$. For dilute suspensions ($c \to o$) and large κa, $H(\kappa a,c) \to 1$ whence [4.14] reduces to Smoluchowski's classic result [1.2], as required.

The predictions presented previously (1) are in qualitative agreement with the rather sparse experimental data available in the literature. Although there are many reported observations concerning the retarded settling velocities of charged particles, there does not appear to be even one comprehensive set of measurements which would permit quantitative verification of the theoretical predictions. Although several workers have described experiments aimed at measuring the retardation effect, most of the published data is incomplete in the sense that not all of the parameters appearing in [4.1] and [4.14] have been reported (λ, a, c or ψ_a frequently being omitted) so that only a semi-quantitative comparison between theory and experiment is possible.

In fact, many workers have not appreciated that the retardation effect could have been caused by electric double layers at all and hence did not measure any of the relevant electrical parameters. There is thus a definite need for a controlled set of sedimentation experiments to enable a proper assessment of existing theories to be made, and this is currently being pursued. Although the retardation effect is unmeasurably small for a single particle, the effect becomes rapidly pronounced in the case of concentrated suspensions with values of U/U_0 as low as 0.35 being reported for very fine particles ($a \simeq 2\mu m$) in 10^{-5} M KCl solution (1). Obviously this retardation effect is of considerable practical importance and needs to be studied further.

V. REFERENCES

1. Levine, S., Neale, G. and Epstein, N., J. Colloid Interface Sci., 57, 424 (1976).

2. Smoluchowski, M., in "Handbuch der Electrizitat und des Magnetismus", Editor L. Graetz, Vol. 2, Leipzig, 1914.

3. Booth, F., J. Chem. Phys., 22, 1956 (1954).

4. Sengupta, M., J. Colloid Interface Sci., 26, 240 (1968).

5. Levine, S. and Neale, G., J. Colloid Interface Sci., 47, 520 (1974).

6. Levine, S. and Neale, G., J. Colloid Interface Sci., 49, 330 (1974).

7. Levine, S., Marriott, J.R., Neale, G. and Epstein, N., J. Colloid Interface Sci., 52, 136 (1975).

8. Levine, S. and O'Brien, R.N., J. Colloid Interface Science, 43, 616 (1973).

9. Wiersema, P.H., Loeb, A.L. and Overbeek, J. Th. G., J. Colloid Interface Sci., 22, 78 (1966).

10. Happel, J. and Brenner, H., "Low Reynolds Number Hydrodynamics", Prentice-Hall, 1965.

NOTE: By inspection of [3.14] and [4.13] in conjunction with [2.27] and [2.23] it will be seen that to determine E and U we need to determine only X, m, A and B (we do not need C and D). These four constants may be determined directly from [3.2],[3.7],[3.11] and the first equation in [3.20] which, when working to order ψ_a, do not involve $v^*(r)$. In other words, to order ψ_a, we need not consider at all the two boundary conditions involving $v^*(r)$, namely [3.6] and the second equation in [3.20].

THERMODYNAMIC STABILITY OF THE TWO-PHASE DISPERSE SYSTEMS

Yuli Glazman

Department of Chemical Engineering
Tufts University
Medford, Massachusetts

Some disperse systems are known to exist for a long time without noticeable changes and it is of interest to learn whether their stability is of a kinetic nature or it has a thermodynamic meaning. Colloidal dispersions are usually regarded as systems unstable in the thermodynamic sense. However, being of paramount importance this problem is actually much more complicated than it seems at first.

The problem of equilibrium state of disperse systems was initially explored on the basis of the phenomenological thermodynamics. When the degree of dispersion becomes higher, the increase of entropy occurs together with the free surface energy increase. Taking this into consideration Rebinder /1/ developed the views of Volmer /2/ on critical emulsions and came to the conclusion that the total free energy of a disperse system may be reduced with decrease of particles radius r as a result of entropy increase if only the value σ of the interfacial tension is low enough. According to Rebinder /1/ the condition of thermodynamic stability of the two-phase disperse systems is expressed by the inequality

$$0 < \sigma < \sigma_m$$

where σ_m is some boundary value of σ which is determined by the commensurate value of the mean energy of thermal (Brownian) motion. At normal temperatures ($T \approx 300^{o}K$) $\sigma_m \approx 0.1$ erg. cm^{-2}.

To establish the possibility of the existence of the two-phase disperse systems at the thermodynamic equilibrium, it is necessary to analyse the dependence of the system's free energy F on the degree of dispersion of the system.

The thermodynamically stable status is realized in any disperse system when a definite distribution of the particles by their dimensions is reached. Strictly speaking it is necessary to calculate the value of the free energy F of the whole ensemble of particles and to find out the distribution corresponding to the minimum of F. It seems that this problem cannot be solved in the general form. Any other approach is inevitably associated with some additional assumptions and their rightfulness is determined by the results obtained.

Instead of determining the distribution of the particles by their dimensions in the thermodynamic equilibrium another approach was used. The infinite great number of monodisperse systems was considered and we calculated which of them has the minimal value of free energy. Then it is assumed that the maximum on the distribution curve corresponds to the dimensions of the particles of the mentioned system.

The entropy of the ideal disperse system consisting of n particles of the dispersed phase and N molecules of the dispersion medium ($N >> n$) may be represented by the relationship

$$\Delta S = k\left(n \ln \frac{N+n}{n} + N \ln \frac{N+n}{N}\right),$$

where ΔS is the entropy and k is the Boltzmann constant.

The expression for the free energy ΔF in the case of spherical particles is

$$\Delta F = E - kT\left(n \ln \frac{N+n}{n} + N \ln \frac{N+n}{N}\right) + 4\pi r^2 \sigma n,$$

where E is the internal energy of the system.

Analysing this equation we came to the conclusion /3/ that the free energy as a function of the degree of dispersion has one extreme point only which is a maximum. This fact that in terms of the mentioned consideration (σ=const and $N>>n$) the dependence of the free energy upon dispersion does not pass through a minimum means that the disperse system is as a matter of principle unstable. The processes of particles aggregation or, on the contrary, the dispersion of particles into molecules should occur in the system depending on the degree of dispersion.

The same problem concerning the disperse systems in the equilibrium state was considered later by Rusanov and others /4/ on the basis of statistical thermodynamics. In the performed investigations the character of the thermodynamic equilibrium - stable or unstable - was also defined basing on the assumption that there is either minimum or maximum in the expression for the free energy of a system.

Using statistical thermodynamics Kligman and Rusanov /5/ recently derived correlations determining the dependence of the extreme radii r_{min} and r_{max} upon the given parameters of the monodisperse system with solid particles. It follows from their investigation that at the fixed parameters the free energy of a system cannot have one of the two possible extremes only - maximum or minimum. Actually, if there is a maximum, then certainly there should be a minimum too. However, the real existence of a thermodynamic equilibrium (either stable or unstable) of the two-phase disperse system is possible in that case only when the corresponding values of r_{min} and r_{max} substantially exceed the dimensions of molecules of the dispersed phase material. It is shown in the mentioned work /5/ that if the value of σ does not depend on r, the stable thermodynamic equilibrium cannot be realized at temperatures which are not too close to the absolute zero and in no case at normal temperatures. This conclusion holds both for ideal and for slightly nonideal systems, that is for the systems in which the interaction between the particles is taken into account and this interaction may be described using the method of the virial expansion. In all these cases the obtained values of r_{min} are smaller than the molecular dimensions.

Lately we also carried out a similar investigation based on the statistical thermodynamics but the procedure of calculations performed was somewhat different. We have come to the same conclusions and in passing we got some additional results.

We will take advantage of the expression for the system's free energy F in which the independence of σ on dispersion, i.e., on the particles radius r or on their number n is assumed from the very first; n and r are connected by the relation,

$$\frac{4}{3}\pi r^3 n = v = const \tag{1}$$

$$F = F_{ex}^{(id)} + F_1(v,T) + 4\pi r^2 \sigma n + F_2(V',T), \tag{2}$$

where $V' = V-v$, and V is the system's total volume.

The first term in (2) is the part of the system's free energy corresponding to the 6n external degrees of freedom of the translation and rotary motion of n non-interacting solid particles with radius r and density ρ.

$F_1(v,T) + 4\pi r^2\sigma n$ is the free energy associated with the interval energy of particles in the dispersion medium ($F_1(v,T)$ is the particles' free energy without surface energy, and $4\pi r^2\sigma n$ is the interfacial free energy). $F_2(V',T)$ is the free energy of dispersion medium of the volume V-v.

The first term in (2) is derived on the basis of the corresponding expression for the statistical integral $Z_{ex}^{(id)}$

$$F_{ex}^{(id)} = -kT\ln Z_{ex}^{(id)} \tag{3}$$

The expression of the statistical integral Z in general case is

$$Z = \frac{1}{n!h^{fn}} \int e^{-\frac{H(p,q)}{kT}} dp\, dq, \tag{4}$$

where H is Hamiltonian function of a system with n particles; p, q is a combination of fn generalized coordinates and fn canonically conjugated with them generalized momenta; f is the number of the freedom degrees of each particle; h is the Planck constant.

The statistical integral $Z_{ex}^{(id)}$ in (3) is characterized by the Hamiltonian function $H_{ex}^{(id)}$ being a sum of kinetic energy of translation and rotary motion of n particles (the index (id) denotes that the particles mutual interaction is not taken into consideration)

$$H_{ex}^{(id)} = H^{tr} + H^{rot} = \sum_{i=1}^{n} \frac{P_{x_i}^2+P_{y_i}^2+P_{z_i}^2}{2m} + \sum_{i=1}^{n} \frac{M_{x_i}^2+M_{y_i}^2+M_{z_i}^2}{2j} \qquad (5)$$

where $P_{x_i}, P_{y_i}, P_{z_i}$ are the impulse components of the i particle in stationary Cartesian system; $M_{x_i}, M_{y_i}, M_{z_i}$ are the components of angular momentum, i.e., the generalized momenta canonically conjugated to the angular coordinates of the i particle turn relative to the axes of the rotary Cartesian system originating in the inertia center of the particle; m and j are the particle mass and its moment of inertia respectively.

On the basis of (4), (5) we get

$$Z_{ex}^{(id)} = \frac{V^n}{n!h^{6n}} (2\pi mkT)^{\frac{3}{2}n} (8\pi^2)^n (2\pi jkT)^{\frac{3}{2}n}, \qquad (6)$$

where the factor V^n is a result of integration with respect to coordinates x_i, y_i, z_i of the inertia centers of all the particles; the factor $(8\pi^2)^n$ is a result of integration with respect to angular coordinates; the factors $(2\pi mkT)^{\frac{3}{2}n}$ and $(2\pi jkT)^{\frac{3}{2}n}$ are a result of integration with respect to impulses which determine $H_{ex}^{(id)}$ (5).

On the basis of (3), (6), taking into account Stirling's formula $n! \approx n\ln\frac{n}{e}$ and the expressions $m = \frac{4}{3}\pi r^3\rho$; $j = \frac{2}{5}mr^2$, the following relation is obtained:

$$F = kTn\ln\{L\frac{n}{Ve}(\rho T)^{-3}r^{-12}\} + F_1(v,T) + 4\pi r^2\sigma n + F_2(V',T), \qquad (7)$$

where

$$L = \frac{1}{8\pi}\frac{h^6}{k^3}\frac{1}{\{\frac{16}{3}(\frac{2}{5})^{\frac{1}{2}}\}^3} \qquad (7')$$

It follows from (1)

$$\frac{dr}{dn} = -\frac{r}{3n}. \qquad (8)$$

Bearing in mind this relation we get from (7)

$$\left(\frac{\partial F}{\partial n}\right)_{T,V,v} = kT\ln\left(\frac{n}{V}LT^{-3}\rho^{-3}r^{-12}\right)+4kT+\frac{1}{3}4\pi r^2\sigma, \tag{9}$$

$$\left(\frac{\partial^2 F}{\partial n^2}\right)_{T,V,v} = \frac{1}{n}\left(5kT-\frac{2}{9}4\pi r^2\sigma\right). \tag{10}$$

Setting the right side of (9) to zero we obtain the equation determining the extreme values r_o of the radius r of the dispersed phase particles

$$kT\ln\left(\frac{n_o}{V}LT^{-3}\rho^{-3}r_o^{-12}\right) + 4kT+\frac{1}{3}4\pi r_o^2\sigma=0, \tag{11}$$

where n_o is the value of n, when $r = r_o$.

The expression for the second derivative of F with respect to n in the extreme point is

$$\left(\frac{\partial^2 F}{\partial n^2}\right)_o = \frac{5kT}{n_o}\left(1 - \frac{8\pi}{45}\frac{\sigma r_o^2}{kT}\right). \tag{10'}$$

The sign of $(\partial^2F/\partial n^2)_o$ is determined by the expression in parentheses of the right side (10'): its positive value corresponds to the minimum of free energy, the negative one - to the maximum.

Substituting L in (11) by its value from (7'), expressing n_o/V by r_o and y (where $y = v/V$) on the basis of (1) and multiplying equation (11) by 2/15kT, we obtain

$$\frac{2}{15}\ln\left(\frac{3}{4}y\,\frac{T^{-3}\rho^{-3}k^{-3}h^6r_o^{-15}}{\left\{\frac{16\pi^3}{3}\left(\frac{2}{5}\right)^{\frac{1}{2}}\right\}^3}\right) + \frac{8}{15} + \frac{8\pi}{45}\frac{\sigma r_o^2}{kT} = 0. \tag{12}$$

If we designate

$$X = \frac{8\pi}{45}\frac{\sigma r_o^2}{kT} \tag{13}$$

and eliminate r_o from (12) by means of (13), then equation (12), as well as equation (11) is transformed into the form (see also /5/)

$$X-\ln X=C \tag{14}$$

where C depends on parameters T,ρ,y,σ only. The value of C may be represented in the form

$$C=\frac{2}{5}\ln\{(\rho/(y^{\frac{1}{3}})\}+\frac{7}{5}\ln T-\ln\sigma-0.9558 \tag{15}$$

if one takes into account the numerical values of the universal constants k and h (ρ and σ are expressed in CGS units).

Equation (14) presented graphically in /5/ has two roots (if $C > 1$). One of them is less than 1 (in /5/ it is designated X_{min}), the other one is over 1 (X_{max}). The first root corresponds to the minimum function F(n) (7), consequently $r_o = r_{min}$; the second corresponds to the maximum of the same function ($r_o = r_{max}$).

This result becomes clear directly from the expression of $(\partial^2F/\partial n^2)_o$ (10'), where X is the second term in parentheses.

When C = 1, equation (14) has one solution only: X = 1, consequently according to (10') $(\partial^2F/\partial n^2)_o = 0$. This corresponds to such a combination of parameters, when there is a point of inflection on the curve F(n) only.

When C < 1, equation (14), and consequently equation (11) as well, do not have any solutions, i.e., the function F(n) does not have extremes (the first derivative of F cannot be equal to zero).

Let us consider now the values of r_{max} and r_{min} under the same parameters of the system and analyze their possible values at normal temperatures. This is of interest both for the processes of spontaneous dispersion and for generalization of the theory when the interaction between the dispersed phase particles is taken into account.

On the basis of (13)

$$r_{max} = \{\frac{45}{8\pi}\frac{kT}{\sigma} X_{max} (C)\}^{\frac{1}{2}} \qquad (16a)$$

$$r_{min} = \{\frac{45}{8\pi}\frac{kT}{\sigma} X_{min} (C)\}^{\frac{1}{2}} \qquad (16b)$$

where X_{max} (C) and X_{min} (C) are the roots of equation (14) under the given value C (C > 1) determined by the expression (15).

As it is seen in tables 1 and 2 (see also the diagram taken from paper /5/), X_{max} increases, and X_{min} decreases rapidly with the increase of C. One can easily draw the inference on the basis of (14) that at high values of C $X_{min} \simeq e^{-C}$.

If T = 290 degrees ($kT \simeq 4 \cdot 10^{-14}$ erg), and $\rho/y^{\frac{1}{3}}$ has a reasonable value of 10 g $\cdot$ cm^{-3}, then formula (15) may be rewritten in form

TABLE 1

Interrelation between X_{max} and C

X_{max}	C	X_{max}	C	X_{max}	C
1	1	6	4.2082	11	8.6021
2	1.3069	7	5.0541	12	9.5115
3	1.9014	8	5.9206	13	10.4350
4	2.6137	9	6.8028	14	11.3609
5	3.3906	10	7.6974	15	13.2274

TABLE 2

Interrelation between X_{min} and C

X_{min}	C	X_{min}	C	X_{min}	C
1	1	0.2	1.8095	10^{-3}	~6.91
0.8	1.0232	0.1	2.4026	10^{-4}	~9.21
0.5	1.1931	0.01	~4.61	10^{-5}	~11.51

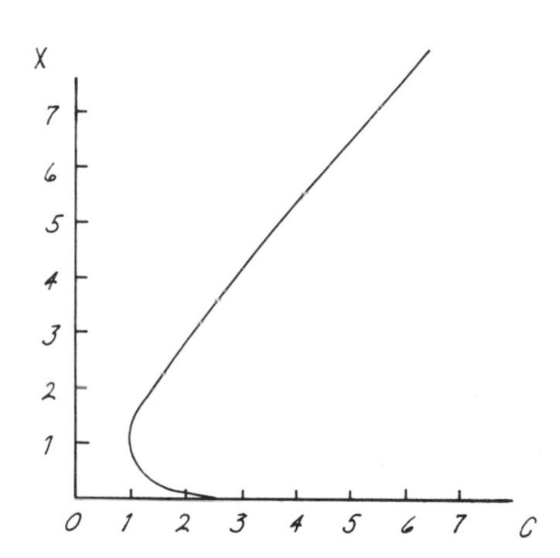

Solution of Equation (14) at Various C Values

$$C = C_o - \ln \sigma \tag{17}$$

where C_o = 7.00. Instead of (16a) and (16b) we get

$$r_{max} \approx 2.68 \left\{\frac{X_{max}(C)}{\sigma}\right\}^{\frac{1}{2}} \cdot 10^{-7} \text{ cm} \tag{18a}$$

$$r_{min} \approx 2.68 \left\{\frac{X_{min}(C)}{\sigma}\right\}^{\frac{1}{2}} \cdot 10^{-7} \text{ cm} \tag{18b}$$

Formulae (18a), (18b), (17) determine the dependence of r_{max} and r_{min} on σ under the above conditions. It is seen from these expressions (see also Tables 1, 2 and the diagram) that when σ is reduced (with C increase), r_{max} increases more rapidly than $\sigma^{-\frac{1}{2}}$; r_{min}, on the contrary, decreases and this decrease is slower than that of $/X_{min}(C)/^{\frac{1}{2}}$.

When σ is small enough, i.e., when the values of C are so high that the relation $X_{min}(C) \approx e^{-C}$ holds, r_{min} tends to lower limit:

$$r_{min} = 2.68 \, (e^{-C_o})^{\frac{1}{2}} \cdot 10^{-7} \text{ cm}; \tag{18b'}$$

TABLE 3

Dependence of r_{max} and r_{min} on σ

(T = 290 degrees, $\rho/y^{1/3} = 10\ g\cdot cm^{-3}$)

σ, $erg\cdot cm^{-2}$	C	X_{max}	X_{min}	r_{max}, cm	r_{min}, cm
~992	1	1	1	$8.50\cdot 10^{-9}$	$8.50\cdot 10^{-9}$
290	2.23	3.48	0.121	$2.94\cdot 10^{-8}$	$5.48\cdot 10^{-9}$
10^2	3.30	4.83	$3.83\cdot 10^{-2}$	$5.93\cdot 10^{-8}$	$5.25\cdot 10^{-9}$
1	7.90	10.2	$3.70\cdot 10^{-4}$	$8.57\cdot 10^{-7}$	$5.14\cdot 10^{-9}$
10^{-2}	12.5	15.2	$3.70\cdot 10^{-6}$	$1.05\cdot 10^{-5}$	$5.14\cdot 10^{-9}$

this follows directly from (18b), (17) (when $C_o = 7.90$, this threshold value is equal to $5.14 \cdot 10^{-9}$ cm). The upper limit of r_{min} is restricted by the requirement that the C value should exceed unity at least slightly. This threshold value is derived (at $C_o = 7.90$) from (17), (18b) assuming C = 1. It can also be determined otherwise - directly from equation (12) at X = 1. The calculations result in a value of $8.50 \cdot 10^{-9}$ cm. Thus

$$5.14 \cdot 10^{-9} \text{ cm} < r_{min} < 8.50 \cdot 10^{-9} \text{ cm}$$

According to (15), (17), (18b') the value of C_o can be reduced, and the values of r_{min} increased respectively at the given temperature (290 degrees) by reducing the first term in (15). However, such a reduction is restricted since the value of ρ cannot be too small (solid particles) and $y^{1/3}$ should not be close to 1. Even provided $\rho/y^{1/3} = 1 \text{g} \cdot \text{cm}^{-3}$, the value of C_o does not reduce much ($C_o \approx 7.0$). In this case the values of r_{min} fall into the range:

$$8.2 \cdot 10^{-9} \text{ cm} < r_{min} < 1.8 \cdot 10^{-8} \text{ cm}.$$

In Table 3 the results of calculation of r_{max} and r_{min} are given for some values of σ and corresponding values of C when $C_o = 7.90$. The above regularities associated with C increase caused by the reduction of σ are reflected there. As it is seen the values of r_{min} (when C_o has a reasonable value) at any possible values of σ are very low (less than the dimensions of molecules) and are restricted by certain limits both on the side of high values of σ and on the side of small ones. The value of r_{max}, on the contrary, increases indefinitely with reduction of σ and can exceed considerably the molecular sizes (if σ is low enough) even in the case of big molecules of the substance of disperse phase. The table illustrates that the function F(n) for the same system and parameters (the same ρ,σ,T,y) has two extremes - the maximum and the minimum. However, while the unstable states are realized the stable equilibrium is impossible (the values of r_{min} are too small). The last assertion may also be true for temperatures which are much lower than the normal temperature (see /5/). As to the spontaneous dispersion, its region is the wider, the smaller the σ value is.

It is worthwhile mentioning that when the limits of the possible r_{min} values have been investigated, the systems were regarded as ideal in that sense only that the interaction between the particles was not taken into consideration in the expressions (2), (7) for the free energy. However, it should be emphasized that the high concentrations of dispersions are not excluded (for example, when $\rho/y^{1/3} = 10\text{g} \cdot \text{cm}^{-3}$ the value of $y^{1/3} = 0.5$ possible, and this corresponds to $y = 1/8$). This means that the inference concerning the small dimensions of r_{min} and accordingly the conclusion about the lack of thermodynamic stability at normal temperatures holds not only for low but also for high concentrations of the dispersed phase (providing only the particles interaction is not taken into account). Such disregard of the particles interaction is certainly the less admissible, the higher the concentration of the disperse system is.

The calculation of the particles interaction and its effect on the free energy of a system is based on the substitution of the Hamiltonian function $H_{ex}^{(id)}$ (5) by the expression

$$H_{ex} = H_{ex}^{(id)} + U(x_1,y_1,z_1,\dots\ x_n,y_n,z_n), \tag{19}$$

where U is the mutual potential energy of all the particles. Instead of the statistical integral $Z_{ex}^{(id)}$ (6) the expression is obtained

$$Z_{ex} = Z_{ex}^{(id)}\ I/V^n \tag{20}$$

where

$$I = \underbrace{\int\dots\int}_{n} e^{-\frac{U}{kT}}\ dx_1,dy_1,dz_1,\ \dots\ dx_n,dy_n,dz_n \tag{21}$$

is the so called configuration integral. Thus an additional term appears in the expression for the system's free energy according to (3), (20); we shall designate it $F_{ex}^{(pot)}$

$$F_{ex}^{(pot)} = -kT\ln I/V^n. \tag{22}$$

$F_{ex}^{(pot)}$ can be calculated by means of approximate methods.

Basing on the method of virial expansion developed for the pressure of nonideal gases (see for instance /6/) an additional term has been introduced in the expression of the disperse systems free energy, and an attempt was made to examine the effect of the particles interaction on the possibility of establishing thermodynamic equilibrium /4/. The first approximation of virial expansion

$$F_{ex}^{(pot)} = kTnB_2\frac{n}{V}, \tag{23}$$

where B_2 is the second virial coefficient, and the proportionality between B_2 and r^3 was used /4,5/. Besides, the condition of applicability of approximation (23)

$$\left|B_2\frac{n}{V}\right| \ll 1 \tag{24}$$

has been taken into consideration /5/. It was established in /5/ that the main results obtained for ideal systems (the conclusion concerning the impossibility of thermodynamic stability at normal temperatures in particular) also hold for slightly nonideal systems.

It has to be noted that strictly speaking the proportion between B_2 and r^3 is correct when the interaction function u(R) determining the value of B_2

$$B_2 = 2\pi \int_0^\infty (1-e^{-\frac{u(R)}{kT}})R^2\, dR \tag{25}$$

depends only on the ratio of the distance R between the centers of two particles to the particle's diameter in the whole region of integration (25) (see also /4/). Therefore, the mentioned condition should be regarded as an admissible approximation for the interaction function. In this case, if one takes into account (1),

$$B_2\frac{n}{V} = B_2^{*}(\frac{v}{V}), \tag{26}$$

where

$$B_2^{*} = 12 \int_0^\infty (1-e^{-\frac{u(\xi)}{kT}})\xi^2 d\xi \quad \text{and} \quad \xi = \frac{R}{2r}. \tag{25'}$$

It is seen from (25') that B_2^* does not depend on dispersion; it is determined by the type of function $u(\xi)$ and by temperature T.

According to (26) $B_2 \frac{n}{V}$ does not depend on dispersion either and it is a function of parameters T and $y = \frac{v}{V}$.

It follows from (23), (25'), (26) that

$$F_{ex}^{(pot)} = kTB_2^{*}\frac{v}{V}n = kTf(T,\frac{v}{V})n, \tag{27}$$

$$(\frac{\partial F_{ex}^{(pot)}}{\partial n})_{T,V,v} = kTB_2^{*}\frac{v}{V} = kTf(T,\frac{v}{V}), \tag{28}$$

$$(\frac{\partial^2 F_{ex}^{(pot)}}{\partial n^2})_{T,V,v} = 0. \tag{29}$$

Since the first derivative of $F_{ex}^{(pot)}$ with respect to n does not depend on dispersion, and the second derivative at any n (or r) including the extreme values (n_o, r_o) is equal to zero, the expressions (10) and (10') are the same independent of the interaction between the particles. An additional term $kTB_2^{*}\frac{v}{V}$ which does not depend on dispersion appears in the expressions (9) and

(11). For this reason the designation (13), the transform of equation (11) into form (14), the equation (14) itself and its solution for a given C do not change when the particles interaction is taken into consideration. The additional term $-\frac{2}{15}B_2^* \frac{v}{V}$ appearing in expression (15), basing on (24), changes the value of C only slightly (see also /5/). Thus it follows that, if the approximation (23) is considered, the interaction affects slightly the values of the particles extreme radii determined by the formulae (16a) and (16b).

Let us consider now the value of $F_{ex}^{(pot)}$ in a more general form than (23) (see also /4/).

$$F_{ex}^{(pot)} = nkT\left(B_2\frac{n}{V} + \frac{B_3}{2}\frac{n^2}{V^2} + \frac{B_4}{3}\frac{n^3}{V^3} + \ldots\right). \qquad (23')$$

In distinction to the approximation (23), the expression (23') - like a virial series for pressure /6/ - takes into account not only the pair interactions but the particles collisions of higher order too (by 3, by 4 and so on).

We will assume that the interaction function of several particles determining an appropriate virial coefficient of series (23') is equal to the sum of pair interaction functions of these particles. It follows from the approximation admitted before for the pair interaction that the proportion exists not only between B_2 and r^3 but also between B_3 and $(r^3)^2$, B_4 and $(r^3)^3$ and so on. Therefore the expression (23') is reduced to the form

$$F_{ex}^{(pot)} = kTn\{B_2^*(T)\frac{v}{V} + B_3^*(T)\left(\frac{v}{V}\right)^2 + B_4^*(T)\left(\frac{v}{V}\right)^3 + \ldots\} = kT\varphi(T,\frac{v}{V})n, \qquad (27')$$

where the expansion in powers of the small parameter v/V is clearly reflected.

We get from (27')

$$\left(\frac{\partial F_{ex}^{(pot)}}{\partial n}\right)_{T,V,v} = kT\varphi(T,\frac{v}{V}), \qquad (28')$$

$$\left(\frac{\partial^2 F_{ex}^{(pot)}}{\partial n^2}\right)_{T,V,v} = 0. \qquad (29')$$

Thus, as in the particular case of virial expansion, the first derivative of $F_{ex}^{(pot)}$ with respect to n does not depend on dispersion, and the second derivative is equal to zero. Consequently, the stable thermodynamic equilibrium cannot arise because of the effect of the particles interaction.

It was admitted formerly /4/ that the second derivative of the free energy with respect to n $(\partial^2F/\partial n^2)_o$ can supposedly change its sign under the action of the particles interaction. Actually neither the sign nor the absolute value of the second derivative of the function F(n) changes at any values of n (including the extreme point), when an additional term $F_{ex}^{(pot)}$ is present in the expression for F(n), since $(\partial^2F_{ex}^{(pot)}/\partial n^2)_{T,V,v} = 0$. The interaction considered in (27') tells only on the appearance of an additional constant value $-\frac{2}{15}\varphi(T,\frac{v}{V})$ in (15). This entails change of the extreme values of the particles radius: reduction of r_{min} and increase of r_{max}, if $\varphi(T,v/V) < 0$, and increase of r_{min} and reduction of r_{max} when $\varphi(T,v/V) > 0$.

In distinction to the slightly nonideal systems described with the first approximation of the virial expansion method, where the condition (24) holds, in the more general case considered here neither the sign nor the absolute magnitude of $\varphi(T,v/V)$ (27') is known at present. Only such a question can be put: what value should have $|\varphi(T,v/V)|$, if $\varphi > 0$, to be sufficient to increase the dimensions of r_{min} from the sizes smaller than the molecular ones to the sizes of the real colloidal particles.

The expression (18b') determining the lower limit of r_{min} holds both for high and for low values of C_o; the only condition is that the value of C in (17) should be big enough (to make the approximation $X_{min}(C) \simeq e^{-C}$ justified).

The additional term $-\frac{2}{15}\varphi$ in (15) can be taken into account by means of substitution of C_o in (17) and in (18b') by value $C_o^{\varphi} = C_o^{\varphi} - \frac{2}{15}\varphi$ where $\varphi = \varphi(T,v/V)$ according to (27'). To increase the limiting value of r_{min} (18b') by a factor of 10 as a result of interaction (27') the following relation should be satisfied

$$\left(\frac{e^{-C_o^{\varphi}}}{e^{-C_o}}\right)^{\frac{1}{2}} \equiv \left[\frac{e^{-(C_o-\frac{2}{15}\varphi)}}{e^{-C_o}}\right]^{\frac{1}{2}} \equiv e^{+\frac{1}{15}\varphi} = 10. \qquad (30)$$

Hence $\varphi = 15 \ln 10 \simeq 34.5$.

To increase the limiting value of r_{min} by a factor of 10^2 it is necessary to have $\varphi \simeq 69$.

As it follows from the most general reasonings such high values of φ are rather unlikely. Let us remind that when the restriction by the first term of virial expansion is admissible (i.e., when the condition making the Van der Waals equation possible is justified)

$$|\varphi(T,\frac{v}{V})| = |f(T,\frac{v}{V})| = |B_2^*\frac{v}{V}| = |B_2\frac{n}{V}| << |. \qquad (31)$$

Thus one has to draw the inference that the achievement of the stable thermodynamic equilibrium of the two-phase disperse systems is hardly possible at normal temperatures even when $\varphi > 0$.

This conclusion certainly does not exhaust the whole problem. The above calculations were performed at σ = const. and another case, when the dependence of σ on r was taken into consideration, had been also explored /7/. It was shown that the stable thermodynamic equilibrium of two-phase disperse systems can be realized as a matter of principle if only two inequalities hold simultaneously.

$$o < \sigma < \sigma_m \qquad (32)$$

$$-\frac{d\ln\sigma}{d\ln r} > 2 \qquad (33)$$

The problem is to find out whether such systems actually exist and what are the real conditions under which these criteria are justified.

REFERENCES

1. Rebinder P.A., Kolloid Zh. (Russ.), 20, 527, 1958; Shchukin E. D., Rebinder P. A., ibid., 20, 645, 1958.
2. Volmer M., Z. Phys. Chem., 125, 151, 1927; A 155, 281, 1931; 206, 181, 1957.
3. Glazman Yu. M., Kolloid Zh. (Russ.), 29, 478, 1967.
4. Rusanov A. I., Kuni F. M., Shchukin E. D., Rebinder P. A., Kolloid Zh. (Russ.), 30, 735, 1968.
5. Kligman F. I., Rusanov A. I., Kolloid Zh. (Russ.), 39, 44, 1977.
6. Landau L. D., Lifshitz E. M., Statistical Physics, 2-nd Ed., 1969; Addison-Wesley Publishing Company.
7. Barboy V. M., Glazman Yu. M., Fuks G. I., Kolloid Zh. (Russ.), 32, 321, 1970; Barboy V. M., Glazman Yu. M., Rebinder P. A., Fuks G. I., Shchukin E. D., Kolloid Zh., 32, 480, 1970.

AUTHOR INDEX

F

G

H

I

J

K

L

M

T

U

V

W

X

Y

Z

SUBJECT INDEX

T

U

V

W

X

Z